Cosmology of Light

A Mathematical Integration of Matter, Life, History & Civilization, Universe, and Self

Pravir Malik, Ph.D.

ISBN-13: 978-0-9903574-7-6

To Sri Aurobindo

Author's Note

This book is about Light. It is a story of Light from an apparent beginning to the present, and then also a story of that condition that transcends Time and is hence beginning-less. But it is more too. It is a book about the central place of Light in all things. Being so, it is a cosmology – it suggests a nature of the universe – in which all that is, is of Light. Hence this book describes a Cosmology of Light.

But the medium chosen to express this nature is mathematics. Specifically, as will be discovered there is a particular symmetrical "function-based" mathematics that is formulated to express the nature of cosmos.

The starting point is Light itself, and all that is seen to exist is as a play of Light. Hence, Light is presumed to exist in some native state characterized by an infinite speed. It chooses or experiments to project its fundamental nature at some reduced speed that is experienced as 'c', 186,000 miles per second in vacuum that has a profound effect on the experienced nature of reality. This projection precipitates a Big Bang. A mathematical state in which quanta becomes a door-way to multiple-layers of Light comes into existence at that moment and is captured by a seed-equation that is positioned as being always fully present in any and everything. Hence all of Light conspires to create anything and everything no matter how different from Light it may seem. Hence there is a fundamental symmetry to everything no matter how different it may appear in form.

The seed-equation is structured by a four-foldness anchored on implicit properties in Light. This four-foldness is the cornerstone of the constructed function-based mathematics. This equation, referred to as the "Light-Space-Time Emergence" equation contains all possibility, and will indicate how all layers of known existence have emerged. The same equation essentially being used for anything in existence points to the fundamental mathematical symmetry of all things.

Hence, from an initial fourfold space-time-energy-gravity expression, through the electromagnetic field, quantum particles, atoms, cells, and then human attributes and dynamics, to complex organizations and countries, culminating in a possible sustainable global civilization, the relationship of all things to this underlying seed-equation and all things as the seed-equation will be emphasized.

But it will also be seen that Love is the compelling dynamic of projected four-foldness and appears to be the magician that causes the separated fourfold powers in Light to reintegrate to make materially real the compelling nature of oneness that Light and all its creatures in reality are.

In the scheme of things, the human, placed at the middle of sixty orders of cosmic magnitude, ranging from the minute Planck-length to astronomical gigaparsecs appears to have a critical role in the continuing unfolding of the Cosmos. It is only by transcending smallness and becoming one with the magician to so enter into the very dynamic of Love, that the universe can continue on its ever-complexifying, ever more-beautiful journey of increasing functional-richness.

It is only if humans so unite with the secret of Light that truer individuality, truer-knowledge, truer-power can emerge, and the very material process of space-time-energy-gravity quantization can be activated to create foundations of super-matter, that in the scheme of things is the compelling crystallization of the infinity in Light, reformulating micro and macro orders of magnitude to change the nature of Cosmos.

In its exploration this book proposes several unique aspects of a cosmology. These include the following:

1. Multiple and simultaneous realities, essentially created through light traveling at different speeds.
2. Fourfold properties of Light, worked out at Light in its native state of infinity, and Light traveling at c, and a one-to-one correspondence between each of the four properties of Light at infinity and Light at c.
3. A unique perspective on a Big Bang as being the result of Light projecting itself from its native state to a state characterized by a slower speed. The Big Bang as presented here is inevitable since the slowdown of Light creates quanta and then matter.

4. The formation of a seed-equation that provides insight into all things as formations of Light.
5. The Light-Matrix of the seed-equation that describes the reality created as Light transforms itself from its native state to that state of traveling at c. The four resulting realities described in the Light-matrix are proposed as being symmetrical hence reinforcing Light to be of the same essence.
6. The Space-Matrix that describes how, given the layers described by the Light-Matrix, the influence of meta-layers can begin to transform the material layer.
7. The notion of four-foldness architecting fundamental constituents of each emergent layer from space-time-gravity-energy, through the electromagnetic spectrum, quantum particles, atoms, cells, and on through to human capacities, organizations, nations, and global sustainable civilization.
8. The notion of space-time-energy-gravity quantization that changes material reality thereby allowing evolution of natural and human history.
9. The notion of quantum certainty precipitating space-time-energy-gravity quantization thereby allowing matter to continue to evolve as a functionally-rich super-matter.
10. The notion of space-quantization being potentially related to dark energy and the expansion of the universe.
11. The notion of humanity initiating a process to change the reality of the "upper" 30 orders of macro-magnitude through consciousness.
12. Self as an all-pervasive single equation that exists in all things, and of which all things are the play.

This then may appear as a culminating statement on a Cosmology of Light – *To be able to change the very nature of the Cosmos, which is nothing other than Light, through the action of conscious fourfold projections of Light that themselves continue to crystallize all the infinite beauty inherent in Light, resulting in a single Love-filled embrace spanning layers of increasingly super-material perpetuity, which is all the time in all aspects just the completeness of Light.*

Pravir Malik,

San Francisco

Previous Books in *Cosmology of Light* Series

- Book 1 – A Story of Light: A Simple Exploration of the Creation and Dynamics of This Universe and Others
- Book 2 – Oceans of Innovation: The Mathematical Heart of Complex Systems
- Book 3 – Emergence: A Mathematical Journey from the Big Bang to Sustainable Global Civilization
- Book 4 – Quantum Certainty: A Mathematics of Natural and Sustainable Human History
- Book 5 – Super-Matter: Functional Richness in an Expanding Universe

SECTION 1: FUNDAMENTAL CONCEPTS

Chapter 1.1 explores the impact that light has on the experienced nature of reality. In particular the impact of light traveling at potentially different speeds will be examined for its ability to create realities based on its speed. Further the Big Bang and a pervasive seed-equation are posited as existing as a result of the slowing down of Light.

Chapter 1.2 explores a theory for the creation of quanta related to the speed of light. Quanta are intimately associated with the seed-equation that acts as a doorway, as it were, into the multi-layered structure represented by the seed-equation. But time, space, and gravity can also be thought of as existing through quantization.

Chapter 1.3 examines some macro-possibilities, such as functional-richness, super-matter, all as being a play of "Light Mathematics" and an expanding and contracting universe set up due to Light.

Chapter 1.1: Light and the Realities It Creates

This chapter will look at the impact light has on creating practical realities.

Light@c, Quanta, and The Reality It Creates

Everything is made from light. But how can this be. How can the dense material objects that make up our world – the stones, cars, buildings – be made from light? How can our thoughts, our emotions, our cells and bodies be made from light? How can all the animals and plants and clouds be made from light? Here is an exploration on now this may be possible.

To begin this exploration think for a minute as to how fast light is traveling. If we turn on a light switch in a room it immediately gets filled with light. If we turn on a light torch on a dark night it immediately lights up a path through the darkness. And yet the light from the sun takes about eight minutes to reach the earth – this speed through a vacuum is referred to as the physical constant 'c'. So while light is moving incredibly fast it is still not moving at the infinite speed we may think it is moving at.

Now, it is because the light is not traveling at an infinite speed that quanta and subsequently the world of matter can appear. All matter is made from atoms. So let us enter into the world of a hypothetical atom in the process of creation to explore how this might be. Imagine that some form of light originates from what will become the nucleus within an atom. By virtue of its finite speed light would take some time, however small, to travel any minute distance within the forming atom. During that fraction of time there is a build up of energy and it is this build up of energy that forms a packet or quanta, as it were, that allows matter to form. In the absence of such a build-up potential-matter would simply disperse. Hence, it is this build up on energy that variously expresses itself as quarks and leptons and bosons and subsequently atoms – the very building blocks of everything that is.

So, in this view, it is because of the finite speed of light, c, that matter is created. That is not to say that matter will be created wherever light is traveling at c, but the possibility exists, so long as other conditions are also fulfilled. But let us leave aside the other conditions and the exact process of such a creation for now.

But what else is implicit or made possible because of the finite speed of light? Well, imagine traveling on a ray of light from the sun to the earth. Imagine that you are in minute 4 of the approximately 8 minute journey. As you look back you will see that 4 minutes in the past you were at the sun. 4 minutes in the future you will be at the earth. And in the present moment you are somewhere between the sun and the earth. So this limited speed of light already creates the concept of time and specifically of the past, the present, and the future.

So, four incredibly fundamental things are created because of the finite speed of light: quanta and therefore matter, the past, the present, and the future. It could be said, therefore, that this matter-based time-bounded universe is a result of the finite speed of light.

Looking at this in another way, it is known too that in a denser medium the speed of light further slows down. Hence, c becomes a fraction of c when traveling through water for instance. So by reversing the logic of such a process this may show that if light slows down from an infinite speed to some other lesser speed such as a multiple or a fraction of c, then the material reality will have to alter, in a similar way as the material reality between vacuum and water is different.

Light @Infinity

But what if light can be experienced at other speeds? What if light can travel at other finite or an infinite speed beyond the c we experience it at, which by the way is an astronomical 186,000 miles per second in a vacuum?

Let us imagine for a minute that light can travel infinitely fast. Think about a big area or volume with a light source at the center. Now since the light travels infinitely fast it will fill up the entire volume instantly. This will be true no matter how large the volume is – it could be the entire solar system, an entire galaxy, or an entire universe. So that light is going to be instantaneously present everywhere – it is going to be omnipresent.

A further thought experiment may show what such omnipresence may look at by considering the night sky. It is because of the finite speed of light that the night sky has only spots of light, however many, across it. But if light were traveling infinitely fast then the night sky would be a canvas or a sea of brilliant white since the light from every corner of the universe, wherever a light source or star existed, would immediately be present. It would appear, in such a thought experiment, that we existed in a sea of light.

Now, since the light is already present everywhere in whatever volume, that is, since that light has already filled up the entire volume, there is nothing else that can similarly arise there that is not of the nature of light. Even if something else were to arise, being surrounded by light it would eventually succumb to that light. So the light is all-powerful or omnipotent within that volume.

Further, since the light exists simultaneously everywhere in that space and has a complete knowledge of itself it therefore has a complete knowledge of that space or of anything that can arise in that space. So it is all-knowing or omniscient in that space.

Finally, the light connects everything together instantaneously and holds these connections and the things connected in its nature, so it is all-nurturing or omninurturing within that space.

So, as a result of the infinite speed of light, light appears to have properties of omnipresence, omnipotence, omniscience, and omninurturing.

The Big Bang & The Seed-Equation

We hear about the Big Bang as the start of the universe. In this Big Bang matter is created. But in the examinations just presented the creation of matter is nothing other than the result of light traveling at the finite speed, and in this universe at 186,000 miles per second in vacuum. So we can say that the Big Bang, the apparent start of the universe, is the result of a slowing down of light from an infinite speed to some finite speed.

When light slows down then energy accumulates in packets or quanta and this results in what was inexpressible being able to express itself as matter. This can be thought of in terms of an incredibly rapidly moving stream of water. If the water is traveling so fast then no boundary will be able to contain it and the energy will be continuous over the length of the stream. If it is traveling slower though, then the water will be able to be held by boundaries along the length of the stream. The energy in this case will be discontinuous and will appear in "packets".

In this process of slowing down the implicit omnipresent-omnipotent-omniscient-omninurturing nature of light becomes or transforms into an explicit or emergent matter-past-present-future nature of light. This process of transformation creates a seed-equation that is present and synonymous with the Big Bang. The dynamics of a resulting universe are contained in the dynamics of such a seed-equation. The seed-equation will be explored in more detail in the section on mathematical foundations.

We have arisen in the field of light that has a finite speed. It is difficult therefore to feel the reality of the omnipresent-omnipotent-omniscient-omninurturing light. We are attuned to perceiving in the material world that has resulted due to the finite speed of light. But if we could step back into the fullness of light in its pure state at infinity from there we would likely see that there can be different universes created as a result of light selectively slowing down to different levels.

The slow down to a different level will create a particular kind of universe.

Note that there have been experiments to slow down light so that it practically moves at a snail's pace. This slowing down has to be put into context. Even if light were to slow down to 1 mile per second, say, from 186,000 miles per second, keep in mind that it is possible that this change in speed is likely only a miniscule fraction of the change in speed from light traveling infinitely fast to light traveling at c, that is, 186,000 miles per second. So in effect what could be said is that when light is made to slow down in experiments, the range or band that it slows down to deviates only slightly from its slower speed in vacuum.

In this view the Big Bang is not an original creative event. It is likely a recurring creative event that allows something innate in light to express itself in time. But what is in light? What is in the omnipresent-omniscient-omnipotent-omninurturing nature of light? Everything we see around us, and an infinite more that we do not see.

Further, quanta being an accumulation of energy or possibility surfacing in the very process of light traveling at c, suggests the unique place it holds in a creation and in entering into anterior or contextual worlds of Light. This will be explored more fully subsequently.

Necessity of Constant Speed of Light and Variable Time and Space

The constant speed of light, at a speed less than an infinite speed, allows a build up of energy at quantum levels, that in turn allows any properties or possibility in light to express itself materially. When light travels at the speed of c – 186,000 miles per second – then the result of that is the material universe as we see and experience it, also with its division of time into a past, a present, and a future. In fact, we could say that for the known universe, as we experience it, light had to travel at c for it to arise. The speed of light had to be constant else there would be a variable, fluid reality to matter, likely displaying barely forming islands subject to sudden disappearance, reappearance, continuing ad infinitum.

But also since the distance traveled by an object or a ray is the result of the speed it is traveling at for the time involved, this gives us an interesting insight that Einstein based his Theory of Relativity on. Since the speed of light is constant, and has to be for the universe to be in its observable stable condition, this means time and distance (or space) potentially can vary. So if an object is traveling very fast, then time and space are going to be experienced differently by it, as compared with an object that is traveling considerably slower: as the speed of an object approaches c time slows down and distance contracts.

So an object that manages to travel at a speed of c will in some sense partake of the reality as experienced by c. It will transcend the conventions of time and space that are set up because of c and experience these differently.

Light at Speed Close to Zero

On the flip-side imagine light traveling at a speed close to or approaching zero. In this case the experienced reality is going to be the opposite of reality as experienced when light is traveling at an infinite speed.

First of all, light emanating from any point will remain fundamentally isolated and become the basis of extreme fragmentation. This has to be since regardless of the amount of elapsed time the light will still be only at its point of origin or source. So instead of a Presence of Light in any considered volume, there will be an Absence of Light.

Further, in imagining an area or volume, such light will have no knowledge of what is going on in the volume, and hence be completely ignorant, will further, have no power to effect any circumstances and hence be completely powerless, and finally, will not be able to coordinate or be present in any relationships. Hence it will be detrimental to or agnostic of any relationships instead of harmonious or nurturing to relationships.

Hence, reality in such a scenario would be stark, dark, desolate, fruitless, completely fragmented, and even perhaps pointless.

Such a reality may represent a negative-infinity and may be thought of as an extreme limit of what may be possible in worlds in which Light projects itself.

Seen from this point of view a Big Bang would be a fortuitous event since it will allow a fruitful reintegration of different worlds, one created by light traveling at an infinite speed, and the other created by light traveling at c, to perhaps begin to take place as will be discussed further.

Chapter 1.2: Light & Quanta

This chapter explores quanta, its relation to light in more detail, and other emergent properties that must also be quantized because of multiple speeds of light.

What is Light & What are Quanta?

When we think of light traveling at the speed of c, there are properties in the reality that emerge as a result of this. These properties as we have examined are of the nature of matter, the past, the present, and the future. These properties are the result of light traveling at c and so can be thought of as emerging from light.

But what do these properties actually mean? And further, how do these properties relate to the apparently quite different properties of omnipresence, omnipotent, omniscience, omninurturing that can appear in a reality where the speed of light is infinite?

Let us consider first the properties of light that emerge when light is traveling at c: the past, the present, the future, and matter.

What is the past? It is the perceivable result of all the work and effort that has taken place so far. It is the foundation upon which the present and future will be built. It represents a status quo, a stability, and even a rigidity, and given that it is the result of the long play of time, it will not easily be persuaded to become another thing. It can be thought of as that which the eye can see when it looks around it. There is "physicality" to what the eye can see and so the essence of the past is a kind of physical-ness. So, ingrained in light, is this ability to project or create physical-ness.

What is the present? It is the tremendous play of forces of all kinds to express themselves here and now. There is "vitality" that is present in this play and often it is the most energetic or forceful of the forces that will win out, as opposed to the most insightful or thoughtful. All the tremendous possibility of the future is seeking for expression now and so this essential vitality can also be thought of as a projection or possibility implicit in light.

What is the future? It is the inevitability of what will manifest. The great thoughts, the great ideas, the purpose, the possibilities will sooner or later express themselves in what we call the future. And the essence of this is thoughtfulness or a curiosity or a purpose that we can summarize as an essential "mentality". So embedded in light is this ability to project mentality.

And what is matter? It is the myriad crystallization into apparent diversity of the one essential reality of light, to allow for a play between these different sides or possibilities in an increasingly harmonious interaction. So its essence is "harmony", and this too can be thought of as an essential property projected from or implicit in light.

But what about when light is considered to move at an infinite speed? Then the properties that become apparent, as already explored, are those of omnipresence, omnipotence, omniscience, and omninurturing. But omnipresence has physical-ness to it, omnipotence has vitality to it, omniscience a mentality to it, and omninurturing a harmony to it. It may be said that physical-ness comes from omnipresence, vitality comes from omnipotence, mentality comes from omniscience, and matter comes from omninurturing.

So whether light is traveling at an infinite speed or the speed we know as c, there is something about the properties it projects, that in essence is the same. So let us refer to these essential properties made apparent through the worlds that are created, as Presence (from omnipresence), Power (from omnipotence), Knowledge (from omniscience), and Harmony (from omninurturing).

Further, quanta can be thought of as existing on that very border of the world or reality created due to Light traveling at c, and realities created as Light travels faster, which at its limit is an infinite speed. So quanta are a doorway or an interface into worlds of Light, and a doorway by which possibilities in deeper worlds of Light can express themselves here in this material realm.

Structured Time, Structured Space, and Their Relation To Quanta

In considering the worlds that are created due to the way light moves or the play of light, we see properties that are projected because of it. In the finite world, the world that results from light traveling at the speed c, there is, relatively, smallness we can grasp and on which we can build. In the infinite world that results from light traveling at an infinite speed there is a vastness and fullness that is difficult to grasp.

The notions of time and space are something entirely different in both worlds, and it could be said that Space allows the full play of everything meant by Power, Knowledge, Harmony, and Presence to be seeded in it, and that Time allows that seeding to flower into fuller forms with its passage.

In other words both space and time are not just abstract concepts but are essentially highly structured to allow physical-ness, vitality, mentality, and matter to become Presence, Power, Knowledge, and Harmony.

Space allows all the possibilities present in Presence, Power, Knowledge, and Harmony and seeded in vast diversity, to evolve into more fullness through the time stages of physical-ness, vitality, mentality, and evolving matter.

Such a creation of space and time is synonymous with the creation of quanta. Quanta become the means for the possibilities inherent in the anterior worlds of Presence, Power, Knowledge, and Harmony to express themselves in a structured space and time. Quanta are therefore a passage into deeper worlds of Light, and a means for possibilities in these deeper realms to express themselves materially.

But further, in this view it may also be proposed that space and time being so structured by the four properties of Light, need also to be quantized. That is, space and time must be experienced as quanta as well.

The Quantization of Space, Time, Matter, and Gravity

Chapter 1.1 proposed that quanta is the result of the slowing down of light, and as such becomes the basis for material expression. In other words, for matter to express itself requires quanta. But further it was also just proposed that space, consisting of seeds of the properties of light, would also require quanta to express itself. Philosophically this has to be in a world where light is traveling at a fraction of its possible speed. Further, it was also proposed that time is highly structured, expressing the growth of the seeds in space through definite phases. As such, growth through such phases can also be thought of as happening due to quantization that so allows phases to express themselves.

But if we take a deeper look at space, time, and matter in light of the properties of Light, it can be deduced that space, consisting of a vast array of seeds derived from the properties of Light is itself an expression of Light's property of Knowledge.

Time, bringing forth the meaning contained in the seeds, regardless of circumstance, and even being opposed by circumstance, can be thought of as Light's property of Power.

Matter itself, being a container in which space and time can allow deeper properties of Light to become materially tangible, must be an expression of Light's property of Presence.

But it is also known from Einstein's General Theory of Relativity that gravity is associated with mass and space, in that it is none other than a mass's instruction telling space how to curve, and again is nothing else that space's instruction telling mass how to move through it. As such, where mass and space exist, there gravity has to exist as well. Hence it may be inferred that gravity is none other than an expression of Light's property of Harmony, which fixes the collective relationship between object and object.

But if as proposed matter, space, and time all need to be quantized in order to express themselves, then this must also be true of gravity.

The section on Mathematical Foundations will begin to explore this in greater detail.

Chapter 1.3: Some Macro-Possibilities Due to Light

This chapter examines some macro-possibilities, such as functional-richness, super-matter, all as being a play of "Light Mathematics" and an expanding and contracting universe set up due to Light.

Functional-Richness, Super-Matter & Expanding Space

Quantization of space, time, energy, and gravity can be thought of as a means by which Light's properties of Knowledge, Power, Presence, and Harmony can manifest materially to so change the very fabric of experienced reality. Quantization happens because light travels at a fraction of the speed with which it can potentially travel. In the process there is a build-up of some aspect of light, depending on the quantization occurring.

The very bases of matter starting from fields, through quantum particles, through atoms and subsequently molecules can all be modeled as emergences of the four-fold properties of Light. In the process of such emergence it would be noticed that each subsequent layer of material complexity contains or embodies greater functional-richness.

Hence it could be proposed that not only is the process of material and subsequent creation driven by over-arching meta-functions, but that the very variance in atoms, and molecules, and subsequently in living cells, for example is also driven by an increasing need to express functional-richness.

If Light contains within it all possibility, then it must be that a creation due to Light exists to fundamentally express all the possibilities contained within it in tangible form.

Space, is proposed as being a seeding ground for infinite variety to emerge. Hence it may also be proposed that as the process of creation is successful so that more and more properties inherent in Light actually get materialized, Space has an incentive to continue to expand to allow more seeds to be 'sowed'. Quantization of space then becomes the means by which that which is inherent in Light seeds itself. As already discussed time-quantization, energy-quantization, and gravity-quantization work with space-quantization to allow such Light-possibilities to manifest.

Since matter is the focal-point of the process of functional-complexification it is conceivable that the trajectory of matter and life experienced so far yields further to a "super-matter" which is just a foundation for an incredible and perhaps unending materialization of functional-richness. In such a view this functional-richness is the possible fate of this therefore expanding universe, so long as certain conditions are fulfilled. Through the use of a function-based mathematics this book is going to explore the conditions that will likely lead to establishing such a fate.

Light Mathematics – Creating the World with Light

When we look at a tree or a flower or an animal there is often a distinctness that becomes apparent. Considering the plant kingdom for example, there is a huge array of medicinal properties that are known to exist in plants. These can in turn be used to address the spectrum of afflictions that surface in human bodies. In other words there is abundant 'function' that exists in the variety of plants. Similarly, without considering the individuality within a type of animal, one can see that the type of animal itself seems to embody a different function. Some animals easily become pets while others will always remain the lone predator. Some function best with distinct roles in a larger collectivity, and some just appear to be wiser.

One can see that there is an intricate combination and play of function that can be behind the million forms we see and experience daily.

In fact imagine the four properties of Light - Power, Knowledge, Service, Harmony as being four massive streams with countless variations within each stream. The variations are consistent with the theme of the overarching property so that sub-properties, in a sense, such as courage, adventure, leadership, justice, upliftment and so on all belong to or exist in the stream of Power. Similarly, wisdom, law-making, subject-specific knowledge, and so on belong to the stream of Knowledge. Perfection, diligence, sincerity, perseverance, humility, and so on along exist in the stream of Presence. Mutuality, compassion, nurturing, relationship, collectivity, and so on, all exist in the stream of Harmony.

Now further imagine that these massive streams rush together and from their union countless seeds are formed. These seeds are such that one of the myriad elements dominates, and then there are possibly numerous other elements that combine with the main one, to create a plethora of unique seeds.

So there is a kind of mathematics of light at work such that the properties and their variations of light become the unique, infinite number of seeds that are the basis of all the infinite functionality and combination of functionality we see around us.

So a plant may be a projection of such a unique functional seed, or may be the result of a number of functions coming together. But since the seed is created in such realms of light, when it is projected into time and space, there are the dynamics of separation from the source that are more dominant, and the seed has to grow and assert what it is in its essence through the play of time and space again. So the inner or light-based creation has to create itself again in worlds apparently separated from light, and in their growth have to reestablish the connection with the larger continent of Light.

So all can be seen as this tremendous 'Light Mathematics' where seeds of uniqueness are created by properties of Light, and then projected into space-time worlds, which as a reminder are a result of the slowing down of Light, and then grow into their essence to establish again a concrete contact with the inner-world of light moving infinitely fast and from which it was created.

An Expanding and Contracting Universe

An expanding universe is the reality of our current universe. But it has also been proposed that the universe will contract at some point, and some have even proposed that such expansion and contraction are recurring and part of a cycle.

We know though that an expanding universe is the result of the Big Bang – that phenomenon when Light slowed down from an infinite speed to the speed of c, in order that some arrangement of possibility contained within it would project or emerge into a matter-based existence. We know further that implicit in the Big Bang and subsequently everything that is created, are the four properties of Light of Harmony, Power, Knowledge, and Service. These are envisioned to be large sets with many nuanced elements in them that combine in infinite ways to create seeds from which all manner of uniqueness emerges in the universe.

In fact, we may say that so long as there continues to be an expansion of the universe this means that these active sets are in fact getting larger. That is, there is a more and more nuanced reality that is being created, which in turn means that the material world is continually opening to the larger worlds of Light behind and always present. And this is the meaning of creation. It is so there is more and more of the possibility in the four properties of Light that surfaces, or emerges, or crystallizes.

In the history of this universe, which will be explored in subsequent sections, we will see that the four essential properties at least emerged as the electromagnetic spectrum, then quantum particles, then atoms, and then also cells. But also we will see that the sets associated with each of the four properties grow astronomically larger as each subsequent layer of the properties of light emerges. For instance, as will be seen, the set of Service at the Cell level, comprised of proteins, is massive since there are over 30,000 proteins in a single human cell.

With each subsequent layer of the emergence of the properties of light these sets have to continue to increase so that the infinite potentiality resident in Light can be realized.

If for some reason though there is a focus on smallness, and myopia, and selfishness, or further separation from fullness, then the four sets cannot increase in size, uniqueness and diversity cannot meaningfully emerge, and the meaning of a creation has been defeated. In this case there is a contraction that begins, and the whole effort will need to wait for some other moment for another Big Bang and another cycle of possibility to emerge.

So expansion, so long as it continues means that the emergent vessels of Light have adapted in such a way that they are allowing more of the nuanced nature of the four properties to be become real and increase the effective size of the four sets. Contraction, on the other hand, means that the emergent vessels of Light are adapting in a way that causes less of the nuanced nature of the four properties to become real as a result of which the field of possibility correspondingly continues to diminish until time and space contract back into the timelessness and spacelessness from which they emerged.

This also gives insight into our universe being a function-based as opposed to a form-based universe. The functions derive from the combination of elements from the four sets that exist in Light. Such "functions" may have been glimpsed in the past as codified by historical mythologies that attribute different functions to different "macro-beings" or gods and goddesses. But these mythologies must have necessarily only glimpsed portions of the infinite sets and so long as we as receptacles of light continue to open to Light there can never be or will never be an end to the range of "functions" that will continue to become apparent.

SECTION 2: THE MATHEMATICAL FOUNDATION

This section explores the mathematics required to interpret the universe as a play of functional-richness. The Light-Matrix describes the overarching structure from a ubiquitous unit to the vast diversity of life. Each layer in the matrix is then explored in more detail through the unit-point, the architectural forces, uniqueness of organizations and its emergence, and the summary inherent dynamics always available. The dynamics of possibility is explored through qualified determinism. Deductively derived mathematical operators that may be used as tools by more conscious organizations will also be explored.

This section will hence, specifically explore the following foundational mathematics:

- Light-Matrix – The Seed-Equation
- Nature of Point-Instant in a System
- Architectural Forces
- Uniqueness of Organizations
- Emergence of Uniqueness
- Inherent Dynamics of Any System
- Qualified Determinism
- Mathematical Operators

Chapter 2.1: Light-Matrix – The Seed-Equation

This chapter will explore the Light-Matrix, a mathematical expression of the codification in Light formed at the time of the Big Bang. This Light-Matrix can be thought of as a seed-equation that provides insight into dynamics of the universe. The Light-Matrix codifies dynamics that may be experienced at the quantum border of realities created through Light traveling at different speeds, and further in the worlds that the quantum veil or window provide access to.

Nature of Light at U

As laid out in Section 1, it is perhaps fair to say that the speed of light has significant implications on the experienced nature of reality. The finiteness, c, at 186,000 miles per second in a vacuum (Perkowitz, 2011), creates an upper bound to the speed with which any object may travel also implying, and as discussed in Section 1, that objective reality will be experienced as a past, a present, a future from the point of view of that object (Einstein, 1995). These characteristics – a past, a present, a future – are implicit in the nature of light and become part of objective reality because of the speed of light. So the way we experience time seems definitely to be determined by c.

Further, c also creates a lower bound when inverted (1/c) being proportional, or arguably even determining Planck's constant, h, that pegs the minimum amount of energy or quanta required for expression at the sub-atomic level. Planck's constant, h, pegging the amount of energy required for expression, therefore may allow matter to form (Lorentz, 1925). Note that Einstein postulated quanta as a fundamental property of light itself, rather than as something that arose in the interaction of light with matter (Isaacson, 2008). Hence, c contributes to or even establishes a reality of nature with a past, present, and future, to also be experienced as a phenomena of connection between seemingly independent islands of matter. This characteristic of 'connection' is therefore also proposed to be implicit in the nature of light and becomes part of objective reality because of the speed of light.

But as explored in Chapter 1.2, a 'past' can also be viewed as established reality as defined by what the eye or other lenses of perception can see. Hence, 'It is the perceivable result of all the work and effort that has taken place so far. It is the foundation upon which the present and future will be built. It represents a status quo, a stability, and even a rigidity, and given that it is the result of the long play of time, it will not easily be persuaded to become another thing. It can be thought of as that which the eye can see when it looks around it. There is "physicality" to what the eye can see and so the essence of the past is a kind of physical-ness. So, ingrained in light, is this ability to project or create physical-ness.'

Also as explored previously, the 'present' is 'the tremendous play of forces of all kinds to express themselves here and now. There is "vitality" that is present in this play and often it is the most energetic or forceful of the forces that will win out, as opposed to the most insightful or thoughtful. All the tremendous possibility of the future is seeking for expression now and so this essential vitality can also be thought of as a projection or possibility implicit in light.'

The 'future' is 'the inevitability of what will manifest. The great thoughts, the great ideas, the purpose, the possibilities will sooner or later express themselves in what we call the future. And the essence of this is thoughtfulness or a curiosity or a purpose that we can summarize as an essential "mentality". So embedded in light is this ability to project mentality.'

These implicit characteristics of the nature of light as experienced at the layer of reality so set up by a finite speed of light may hence be summarized by Equation 2.1.1, where c_U refers to the speed of light of 186,000 miles per second, that has created the perceived nature of reality, U, as expressed in Equation 2.1.1:

$$c_U\text{:}[Physical, Vital, Mental, Connection]$$

Eq. 2.1.1: Implicit Characteristics of the Nature of Light at U

Nature of Light at ∞

Exploring further, it is known however that at quantum levels the nature of reality is characterized by wave-particle duality. Light itself (Feynman, 1985) and matter (De Broglie, 1929) may be experienced as both particles and waves. But for matter to be experienced as waves implies that ‘h’ must have become a fraction of itself, $h_{fraction}$, to allow the concentration or possibility of quanta to have dispersed into wave-form. This further implies that c must have become greater than itself, c_N, such that the inequality specified by Equation 2.1.2 holds:

$$c_N > c_U$$

Eq. 2.1.2: Layer N, Layer U Inequality of Speed of Light

Note that what is implied here is that just as there is a nature of reality specified by U that is the result of the speed of light being 186,000 miles per second, so too there is another nature of reality specified by N that is the result of a speed of light greater than 186,000 miles per second.

This is akin to recent developments in physics with the notion of property spaces being separate from but influencing physical space as explored by Nobel Physicist Frank Wilczek (Wilczek, 2016). But further in “Slow Light” Perkowitz’s recent treatment of today’s breakthroughs in the science of light (Perkowitz, 2011) he states: “Although relativity implies that it’s impossible to accelerate an object to the speed of light, the theory may not disallow particles already moving at speed c or greater.”

So light traveling at c_N may be possible. Current instrumentation, experience, and normal modes of thinking though, having developed as a bi-product of the characteristics so created in the layer of reality U may be inadequate to access N without appropriate modification. The notion of wave-particle duality already challenges the notion of normal thinking perhaps because wave-like phenomena could be a function of faster than c motion, and particle-like phenomena a function of less than or equal to c motion. That these may be happening simultaneously is reinforced by principles such as complementarity in which experimental observation may allow measurement of one or another but not of both (Whitaker, 2006).

But then taking this trend of a possible increase in the speed of light to its limit, this will result in a speed of light of infinite miles per second. The question is, what is the nature of reality when light is traveling at infinite miles per second? As explored in Section 1, in any space-time continuum light originating at any point will instantaneously have arrived at every other point. Hence light will have a full and immediate *presence* in that space-time continuum. Further, that light will *know* everything that is happening in that space-time completely instantaneously – that is know what is emerging, what is changing, what is diminishing, what may be connected to what, and so on - or have a quality of knowledge. It will connect every object in that space-time completely and therefore have a quality of connection or *harmony*. Finally nothing will be able to resist it or set up a separate reality that excludes it and hence it will have a quality of *power*.

These implicit characteristics of the nature of light as experienced at the layer of reality so set up by an infinite speed of light may hence be summarized by Equation 2.1.3, where c_∞ refers to the speed of light of ∞ miles per second, that has created the perceived nature of reality, ∞:

$$c_\infty:[Presence, Power, Knowledge, Harmony]$$

Eq. 2.1.3: Implicit Characteristics of Nature of Light at ∞ Speed

Transformation from ∞ to U

But by (2.1.3) it can also be noticed that 'physical' is related to Presence, 'vital' is related to Power, 'mental' is related to Knowledge, and 'connection' is related to Harmony.

The question then, is how do these apparent qualities at ∞ precipitate or become the physical-vital-mental-connection based diversity experienced at U? This may be achieved through the intervention or action of a couple of mathematical transformations acting on the implicit characteristics of nature of light at ∞ speed as summarized by (2.1.3).

First, the essential characteristics of Presence, Power, Knowledge, Harmony that it is posited exist at every point-instant by virtue of the ubiquity of light at ∞ will need to be expressed as sets with up to infinite elements. Second, elements in these sets will need to combine together in potentially infinite ways to create a myriad of seeds or signatures that then become the source of the immense diversity experienced at U. This suggests that all that is seen and experienced at U may be nothing other than 'information' or 'content' of light and as such that there are fundamental mathematical symmetries at play where everything at U is essentially the same thing that exists at ∞.

Assuming that the first transformation occurs at a layer of reality K where the speed of light is c_K, such that $c_U < c_K < c_\infty$, this may be expressed by Equation 2.1.4:

$$c_K{:}[S_{Pr}, S_{Po}, S_K, S_H]$$

Eq. 2.1.4: The First Transformation at Layer K

S_{Pr} signifies 'Set of Presence', S_{Po} signifies 'Set of Power', S_K signifies 'Set of Knowledge', S_H signifies 'Set of Harmony/Nurturing' (note: Harmony and Nurturing will be used interchangeably through this book).

Assuming that the second transformation occurs at a layer of reality N where the speed of light is c_N, such that $c_U < c_N < c_K < c_\infty$, this may be expressed by Equation 2.1.5:

$$c_N{:}\ f(S_{Pr}\ x\ S_{Po}\ x\ S_K\ x\ S_H)$$

Eq. 2.1.5: The Second Transformation at Layer N

The unique seeds are therefore a function, f, of some unique combination of the elements in the four sets S_{Pr}, S_{Po}, S_K, S_H.

The relationship between the layers of light may be modeled by the following matrix in Equation 2.1.6:

$$Light_{Matrix} = \left| \begin{array}{c} c_\infty: [Pr, Po, K, H] \\ (\downarrow R_{c_K} = f(R_{c_\infty})) \\ c_K: [S_{Pr}, S_{Po}, S_K, S_H] \\ (\downarrow R_{c_N} = f(R_{c_K})) \\ c_N: f(S_{Pr}\, x\, S_{Po}\, x\, S_K\, x\, S_H) \\ (\downarrow R_{c_U} = f(R_{c_N})) \\ c_U: [P, V, M, C] \end{array} \right|$$

Eq. 2.1.6: Light-Matrix

The matrix should be read from the top row down to the bottom row as indicated by the ↓ between rows, and suggests a series of transformations leading from the ubiquitous nature of light implicit in a point – presence, power, knowledge, harmony - to the seeming diversity of matter observed at the layer of reality U which is fundamentally the same presence, power, knowledge, and harmony projected into another form of itself.

The first transformation is summarized by Equation 2.1.7:

$$R_{c_K} = f(R_{c_\infty})$$

Eq. 2.1.7: Light-Matrix First Transformation

This is suggesting that the reality at the layer specified by the speed of light c_K, R_{c_K} is a function of the reality at the layer specified by the speed of light c_∞. This transformation translates the essential nature of a point into the sets described in Equation 2.1.4.

The second transformation is summarized by Equation 2.1.8:

$$R_{c_N} = f(R_{c_K})$$

Eq. 2.1.8: Light-Matrix Second Transformation

This is suggesting that the reality at the layer specified by the speed of light c_N, R_{c_N} is a function of the reality at the layer specified by the speed of light c_K. This transformation combines elements of the sets into unique seeds as suggested by Equation 2.1.5.

The third transformation is summarized by Equation 2.1.9:

$$R_{c_U} = f\left(R_{c_N}\right)$$

Eq. 2.1.9: Light-Matrix Third Transformation

This is suggesting that the reality at the layer specified by the speed of light c_U, R_{c_U} is a function of the reality at the layer specified by the speed of light c_N. This transformation builds on the unique seeds suggested by (2.1.5) to create the diversity of U as specified by (2.1.1).

In this framework the notion of wave-particle duality hence may become complementary block-field-wave-particle "quadrality" where block refers to phenomenon resident to ∞, field to phenomenon resident to N, wave to phenomenon resident to K, and particle to phenomenon resident to U. The block is all the reality always present behind the surface and is captured by (2.1.3) or the top line in the Light-Matrix as expressed in (2.1.6). The field is captured by (2.1.4) or the second line from the top in (2.1.6) and can be thought of as layers of possibility existing in each of the sets. The wave is captured by the creation of seeds represented by (2.1.5) or by the third line in (2.1.6). The particle that is apparently disconnected from the whole is captured by (2.1.1) or the bottom line in (2.1.6).

The block is all the reality always present behind the surface and is captured by (2.1.3) or the top line in the Light-Matrix as expressed in (2.1.6). The field is captured by (2.1.4) or the second line from the top in (2.1.6) and can be thought of as layers of possibility existing in each of the sets. The wave is captured by the creation of seeds represented by (2.1.5) or by the third line in (2.1.6). The particle that is apparently disconnected from the whole is captured by (2.1.1) or the bottom line in (2.1.6).

So what we arrive at is a fundamental Light-Matrix that suggests key dynamics for different layers of Light. Quanta, being at the interface of at least Layer U and Layer N are so positioned as being subject to multiple sets of dynamics.

Layer Zero (0)

For the sake of completeness the thought-experiment to do with light traveling at a speed of zero must now be brought up. In Chapter 1.1 a possible reality that would result if light were to travel at this speed was described as opposite to the reality were light to travel at an infinite speed.

Hence, ubiquitous Presence of Light would become the ubiquitous Absence of Light or Darkness. The aspect of Power would become utter Weakness. The aspect of Knowledge would become complete Ignorance. The aspect of Harmony would become total Chaos.

And all this because Light would be unable to travel form where it is, in complete opposition to its known nature of traveling at a speed of c, and perhaps even at a speed of ∞. Light, hence, would be hidden in itself, so as to speak, in some sort of a negative infinity. An Equation, 2.1.10, Nature of Light at Speed Zero (0), would hence be:

$$c_0{:}[Darkness, Weakness, Ignorance, Chaos]$$

Eq. 2.1.10: Nature of Light at Speed Zero (0)

Further, (2.1.6), the Light-Matrix, would be modified to become Equation 2.1.11, Light-Matrix with Zero-Limit, where c_0 implies light at zero-speed, and D, W,I, and C imply Darkness, Weakness, Ignorance, and Chaos respectively:

$$
Light_{Matrix}(0_{limit}) = \left| \begin{array}{c} c_{\infty}\text{:}[Pr, Po, K, H] \\ (\downarrow R_{C_K} = f(R_{C_{\infty}})) \\ c_K\text{: } [S_{Pr}, S_{Po}, S_K, S_H] \\ (\downarrow R_{C_N} = f(R_{C_K})) \\ c_{N}\text{: } f(S_{Pr}\ x\ S_{Po}\ x\ S_K\ x\ S_H) \\ (\downarrow R_{C_U} = f(R_{C_N})) \\ c_U\text{: } [P,V,M,C] \\ \Uparrow \\ c_{0}\text{:}[D, W, I, C] \end{array} \right|
$$

Eq. 2.1.11: Light-Matrix with Zero-Limit

The implication of the bottom zero-limit line is that just as there is proposed to be an influence from the upper layers of light on U, as will be explored in subsequent chapters, so too there is a subtle influence from this lower layer of "light" that perhaps creates a certain obstinacy of the untransformed (U) nature of practical reality. Such obstinacy will be further explored in Chapters 2.5, 2.6, and 2.7 and perhaps reinforces the need for the notion of 'transformation' in the first place.

In subsequent chapters the Light-Matrix will assume the form as depicted in (2.1.6) as opposed to (2.1.11). The influence of the zero-limit will however be supposed in the operation of Layer U, which therefore, as just proposed will require the necessity of a continuous transformation in order to allow increasing functional-richness to continue. Such an assumption is consistent with observation and as will be explored therefore requires human faculties such as will, thought, love, amongst others in order to counter the influence of darkness, ignorance, weakness, and chaos.

Chapter 2.2: Nature of a Point-Instant in a System

The 'point-instant' captures the inherent innovation that appears to exist in the system and is represented by the top-line in Equation (2.1.6), the Light-Matrix. This innovation is a function of the properties of light derived in Equation (2.1,3) - Presence, Power, Knowledge, and Harmony – and reproduced here for convenience:

$$c_{\infty}\text{:}[Presence, Power, Knowledge, Harmony]$$

The 'point' aspect of the point-instant suggests the space-dimension and that space is seeded with the possibilities inherent in the properties of Presence, Power, Knowledge, and Harmony. The 'instant' aspect suggests the time-dimension and gives insight into the process of emergence that the possibilities in space progressively surface as.

In its point-instant, Presence-Power-Knowledge-Harmony wholeness, Presence allows emergences to continue to develop as per the possibilities implicit in the past-present-future or physical-vital-mental pathway.

Beginning to translate this into an equation, the notation $System_{P_I}$ is given to system-presence. This system-presence is true across any considered Time-Space continuum starting from a time-space boundary '0' to a time-space boundary 'N'. This notion is characterized by the notation $TS_{0\rightarrow N}$. Within that boundary from 0 to 'N', the 'presence' is such that it will always seize an opportunity to cause a shift from the physical-leading to the vital-leading, and from the vital-leading to the mental-leading. Research shows (Malik, 2009) that greater degrees of freedom is afforded by such a shift.

The notion that the presence seizes on opportunity as characterized by the notation:

$$\begin{array}{c} Presence \\ \downarrow \\ Opportunity \end{array}$$

The shift from physical-leading to vital-leading and vital-leading to mental-leading is characterized by:

$$\begin{matrix} P_L & \rightarrow & V_L \\ V_L & \rightarrow & M_L \end{matrix}$$

Hence in this approach it is suggested that:

$$System_{Pr} \equiv TS_{0 \rightarrow N} \left[\begin{matrix} Presence \\ \downarrow \\ Opportunity \end{matrix} \begin{bmatrix} P_L & \rightarrow & V_L \\ V_L & \rightarrow & M_L \end{bmatrix} \right]$$

But there is something else about this Presence as well. All other developments take place in it. That is, it provides a container of sorts in which the plays of system-power, system-knowledge, and system-harmony/system-nurturing can take place. This notion is summarized by the notation:

$$Container_{\begin{bmatrix} System_P \\ System_K \\ System_N \end{bmatrix}}$$

Hence, combining these various components, an equation for 'system-presence', Equation 2.2.1, arises:

$$\boldsymbol{System_{Pr}} \equiv \boldsymbol{TS_{0 \rightarrow N}} \left[\begin{matrix} \boldsymbol{Presence} \\ \downarrow \\ \boldsymbol{Opportunity} \end{matrix} \begin{bmatrix} P_L & \rightarrow & V_L \\ V_L & \rightarrow & M_L \end{bmatrix} \,\&\, \boldsymbol{Container}_{\begin{bmatrix} System_P \\ System_K \\ System_N \end{bmatrix}} \right]$$

Eq 2.2.1: System Presence

In its point-instant, Presence-Power-Knowledge-Harmony wholeness, Power allows emergences to continue to happen in spite of tremendous oppositions of all kinds; this too, regardless of field or area.

Constructing an equation for system-power, the notation $System_P$ is used to represent system-power. Any endeavor will always be met with resistances of various kinds. The resistances that arise along the physical dimension are referred to as P_R. The resistances that arise along the vital dimension are referred to as V_R. The resistances that arise along the mental dimension are referred to as M_R. In the fruition of any endeavor one or all of these types of resistances may arise. Further, resistance of one kind often feeds on resistance of another kind, and to generalize the resistances encountered in an endeavor may be characterized as the product of the three types of resistance

$$P_R * V_R * M_R$$

These resistances arise across any considered Time-Space boundary from 0 to 'N', and therefore it may be said that the power of the system is such that:

$$power > \sum_{TS=0}^{N} P_R * V_R * M_R$$

An equation for 'system-power', Equation 2.2.2, hence, is the following:

$$System_P \equiv power > \sum_{TS=0}^{N} P_R * V_R * M_R$$

Eq 2.2.2: System Power

In its point-instant, Presence-Power-Knowledge-Harmony wholeness, Knowledge orchestrates emergences to continue to happen by leveraging the right instruments and circumstances.

Translating this into an equation, the notation, $System_K$, is used for system-knowledge. This $System_K$ is such that it leverages the right instrumentation and circumstance to bring about the progress that is possible. This concept of 'instrumentation' is denoted by the subscript 'I'. The concept of 'circumstance' is denoted by the subscript 'C'. Both instrumentation and circumstance can be of a physical, vital, or mental type and this possibility is denoted by:

$$\begin{bmatrix} P_{I,C} \\ V_{I,C} \\ M_{I,C} \end{bmatrix}$$

Further, the notion that the 'knowledge' is such that it 'leverages' the right instrumentation and circumstance is depicted by:

$$\begin{matrix} Knowledge \\ \downarrow \\ Leverage \end{matrix}$$

This act of leveraging results in a fundamental shift so that the physical-leading yields to the vital-leading, and the vital-leading yields to the mental-leading. Hence:

$$\begin{matrix} Knowledge \\ \downarrow \\ Leverage \end{matrix} \begin{bmatrix} P_{I,C} \\ V_{I,C} \\ M_{I,C} \end{bmatrix} \rightarrow \begin{bmatrix} P_L & \rightarrow & V_L \\ V_L & \rightarrow & M_L \end{bmatrix}$$

Since this behavior may exist across any Time-Space continuum an equation for system-knowledge, Equation 2.2.3, is suggested:

$$System_K \equiv TS_{0\rightarrow N} \left[\begin{matrix} Knowledge \\ \downarrow \\ Leverage \end{matrix} \begin{bmatrix} P_{I,C} \\ V_{I,C} \\ M_{I,C} \end{bmatrix} \rightarrow \begin{bmatrix} P_L & \rightarrow & V_L \\ V_L & \rightarrow & M_L \end{bmatrix} \right]$$

Eq 2.2.3: System Knowledge

In its point-instant, Presence-Power-Knowledge-Harmony wholeness, Harmony or nurturing allows emergences to continue to happen with more and more degrees of freedom to come to the surface.

This characteristic of this implicit-nurturing may be referred to as 'system-nurturing'. Like the other characteristics it is suggested to exist across a Time-Space continuum. This is depicted by:

$$TS_{0\rightarrow N}$$

There is an action of nurturing such that any state is always advanced to a higher level. This is depicted by:

$$\coprod_{Nurturing} \begin{pmatrix} P_- & M_+ \\ V_- & V_+ \\ M_- & P_+ \end{pmatrix}$$

Hence, there is a 'union', depicted by 'U' that 'nurtures' the negatives towards their positives.

Further, there is an increasing action of nurturing such that the possibility of integration is always increased to form a larger and larger basis. This increasing basis is depicted as being modulated by the polar coordinates 'r' and 'θ', where r is the radius which increases from an initial value of '0', and 'θ' is an angle from '0' to '360'.

Hence, the equation of system-nurturing, Equation 2.2.4, is depicted as:

$$System_N \equiv TS_{0 \to N}\left(\coprod_{Nurturing} \begin{pmatrix} P_- & M_+ \\ V_- & V_+ \\ M_- & P_+ \end{pmatrix} mod\,(r, \theta) \right)$$

Eq 2.2.4: System Nurturing

It is suggested that these four characteristics exist across any system, and to denote this it is generalized that every point in any system is embedded with this four-fold intelligence. It is suggested that this four-fold intelligence is resident in every instant-spot of the system. It is suggested that to be able to leverage or activate this four-fold intelligence at will is the ultimate act of innovation.

Such a depiction of models, as Bar-Yam states in "From Big Data to Important Information" (Bar-Yam, 2016), 'is "valid" only because of the irrelevance of details...If we want to say anything meaningful about a system—meaningful in the sense of scientific replicability or in terms of utility of knowledge—the only description that is important is one that has universality, that is, is independent of details. There is no utility to information that is only true in a particular instance. Thus, all of scientific inquiry should be understood as an inquiry into universality—the determination of the degree to which information is general or specific.'

Chapter 2.3: Architectural Forces

The characteristics embedded in a point as in (2.1.3) suggest a possibility that is hard to fathom. One can only glimpse the extraordinary nature embedded in a point. And yet it can be suggested that this extraordinary nature is barely visible unless the right analytical lens of the sort being suggested here is first set up. Further, it is suggested that this extraordinary nature is responsible for a broader set of architectural forces that exist behind the visible face of things.

Hence, system-presence, system-power, system-knowledge, and system-nurturing that define the nature of every point in our system, become more tangible as a broader set of architectural forces that emanate from each of them.

Considering system-presence, here is a characteristic that appears to be everywhere (Malik, 2009) at the service of all the constructs that develop within it. There is a diligence and perseverance by which any opportunity for progress is seized. Further, if one considers the extraordinary detail that appears in any construct, whether an atom, a body, a planet, or a galaxy, one is struck by the high degree of perfection that surfaces in this presence.

So if one contemplates the nature of this system-presence there is a set of forces that surface. Depicting such a set as $S_{System_{Pr}}$, one can arrive at elements such as Service, Perfection, Diligence, Perseverance, amongst others, that are part of this set. Hence, the set can be described by Equation 2.3.1:

$$S_{System_{Pr}} \ni [Service, Perfection, Diligence, Perseverance, \ldots]$$

Eq 2.3.1: Set of System Presence

Similarly, considering the characteristic of system-power, one can hypothesize that there is a family of forces that emanates from it. The kinds of forces may be thought of as Power, Courage, Adventure, Justice, amongst others. The set for system-power can hence be depicted by Equation 2.3.2:

$$S_{System_P} \ni [Power, Courage, Adventure, Justice, ...]$$

Eq 2.3.2: Set of System Power

Similarly, considering the system-knowledge as the root of various powers that emanate from it, one may characterize the set for system-knowledge by Equation 2.3.3:

$$S_{System_K} \ni [Wisdom, Law\ Making, Spread\ of\ Knowledge...]$$

Eq 2.3.3: Set of System Knowledge

The set for system-nurturing is depicted by Equation 2.3.4:

$$S_{System_N} \ni [Love, Compassion, Harmony, Relationship\ ...]$$

Eq 2.3.4: Set of System Nurturing

Chapter 2.4 describes how unique signatures or seeds for any type of organization can be built by leveraging the elements in the sets of architectural forces.

Chapter 2.4: Uniqueness of Organizations

The hypothesis is that every organization, whether an atom, cell, person, team, corporation, market, or country is unique and that this uniqueness can be specified in terms of elements of the derived sets for power, knowledge, presence, and nurturing. This hypothesis for uniqueness stems from observations at multiple levels.

At the sub-atomic level Nobel Laureate Wolfgang Pauli's 'Pauli Exclusion Principle' states that no two similar fermions, which include fundamental particles with half-integer spin such as protons, neutrons, and electrons, can occupy the same quantum states simultaneously (Pauli, 1964). Spin has to do with the angle that the particle has to rotate through before being symmetrical with its original state. Half-integer spin particles need to rotate through 720 degrees before being symmetrical with their original state. The implication of the Pauli Exclusion Principle is that fundamental structure and consequently stability comes into being at the atomic level, which as is evident in the periodic table allows the separation of function related to form. This stability related to the underlying structure of atoms implies the basis of uniqueness and diversity. In the absence of the Exclusion Principle matter would just be a dense soup (Hawking, 1988) with particles occupying overlapping space.

At the observable level uniqueness is evident from the immense diversity of distinct species on earth (Mora, 2011) estimated to be over 2 million, and further the uniqueness of every member of each species. This member-level uniqueness is suggested by the difference in non-coding regions of the DNA that may vary in their sequence by about 1 to 4 percent, which in turn result in unique protein binding sequences of each human (Snyder, 2010), as an example, which in turn results in unique observable qualities.

At the astronomical level Einstein's Special Theory of Relativity (Einstein, 1995) suggests that every coordinate system potentially has its own space-time rendering as opposed to there being one absolute space and time. This implies the notion of uniqueness as an implicit property of space.

The four properties explored in the Chapter 2.1 and 2.2 define the source of that innovation. From this source emanate 4 sets of forces that suggest the boundaries of that innovation.

Assuming then that the fount of uniqueness is system-presence, a general equation for organizations that belong to the family of system-presence can be derived. Such uniqueness can be depicted as Sig_x where the subscript 'x' refers to the source family, and 'Sig' or signature to 'uniqueness'. Hence the uniqueness of an organization in the family of system-presence would be notated by $Sig_{System_{Pr}}$.

In line with the development of properties of a point and the precipitating architectural forces as discussed in Chapter 2.1 and 2.2 respectively, an approach to constructing such uniqueness is to assume a primary factor X that drives the uniqueness that belongs to the set $S_{System_{Pr}}$. Further, assume that the uniqueness is qualified by a number of secondary factors Y that may belong to any of the 4 sets - $S_{System_{Pr}}, S_{System_P}, S_{System_K}, S_{System_N}$. The primary factor X would have a greater weightage than any of the secondary factors Y. The weightage of X hence could be depicted by the number 'a', and the weightage of Y a number $'b_{0-n}'$, such that a > b. Further, the secondary element can repeat from '0 – n' times, and is hence depicted as $Y\bar{b}_{0-n}$.

The equation, Equation 2.4.1, hence for a unique organization derived from the family of system-presence is:

$$Sig_P = Xa + \overline{Yb_{0-n}} \quad where \begin{bmatrix} X \in [S_{System_{Pr}}] \\ Y \in [S_{System_{Pr}}, S_{System_P}, S_{System_K}, S_{System_N}] \\ a, b \text{ are integers}; a > b \end{bmatrix}$$

Eq 2.4.1: System Presence Based Unique Organization

Similarly, an equation, Equation 2.4.2, for a unique organization derived from the family of system-power is:

$$Sig_V = Xa + \overline{Yb_{0-n}} \quad where \begin{bmatrix} X \in [S_{System_P}] \\ Y \in [S_{System_{Pr}}, S_{System_P}, S_{System_K}, S_{System_N}] \\ a, b \text{ are integers}; a > b \end{bmatrix}$$

Eq 2.4.2: System Power Based Unique Organization

An equation, Equation 2.4.3, for a unique organization derived from the family of system-knowledge is:

$$Sig_M = Xa + \overline{Yb_{0-n}} \quad where \begin{bmatrix} X \in [S_{System_K}] \\ Y \in [S_{System_{Pr}}, S_{System_P}, S_{System_K}, S_{System_N}] \\ a, b \text{ are integers}; a > b \end{bmatrix}$$

Eq 2.4.3: System Knowledge Based Unique Organization

An equation, Equation 2.4.4, for a unique organization derived from the family of system-nurturing is:

$$Sig_I = Xa + \overline{Yb_{0-n}} \quad where \begin{bmatrix} X \in [S_{System_N}] \\ Y \in [S_{System_{Pr}}, S_{System_P}, S_{System_K}, S_{System_N}] \\ a, b \text{ are integers}; a > b \end{bmatrix}$$

Eq 2.4.4: System Nurturing Based Unique Organization

The four preceding equations can be generalized by Equation 2.4.5:

$$Sig = Xa + \overline{Yb_{0-n}} \quad where \begin{bmatrix} X \in [S_{System_{Pr}}, S_{System_P}, S_{System_K}, S_{System_N}] \\ Y \in [S_{System_{Pr}}, S_{System_P}, S_{System_K}, S_{System_N}] \\ a, b \text{ are integers}; a > b \end{bmatrix}$$

Eq 2.4.5: Generalized Equation for Unique Organization

Having considered the structure of uniqueness, the next question is how does such uniqueness emerge? This is discussed in Chapter 2.5.

Chapter 2.5: Emergence of Uniqueness

While the uniqueness of organizations as represented by the Signature is a seed, like any seed there is a process for its emergence (Kaufmann, 1995; Portugali, 2012; Yates, 2012), and the uniqueness will often be hidden or very much behind the scene until certain conditions are fulfilled (Malik, 2009).

The implicit nature of Time and Space suggest a universal developmental model that provides a cue as to the process for emergence (Deep Order Mathematics Videos, 2016). In this model the four sets of architectural forces already described form a pool in space, as it were, from which possibility arises. Possibility itself is unique from point to point and is governed by the Equation for Uniqueness (Equations 2.4.1 through 2.4.5) described in the previous chapter.

Hence it is observed that initially the uniqueness takes a 'physical' form, moving on to a 'vital' form, and then onto a 'mental' form. Once the orientations implicit in each of these phases are assimilated, then the uniqueness takes on an 'integral' form. The integral form is a threshold phase, and allows the uniqueness suggested by the Signature to emerge in fuller force or in its 'force' form. The final phase is the 'contextual form' that allows the signature to act with impunity within a considered context.

Mathematically, if an organization exists at the physical phase, it may be suggested that its signature or uniqueness is modulated by the constant 'π'. π is the seed of a circle or sphere and can be thought of as defining behavior that is tightly bound. Within such a tightly bound volume it will likely not even be apparent what the uniqueness of an organization necessarily is. Assuming the uniqueness to be defined by the derived question *Sig*, the physical-level (P) behavior can be described by the following equation-segment where 'mod' signifies modulated-by:

$$P\text{: } Sig * mod\ (\pi)$$

If an organization exists at the vital level, it may be suggested that its uniqueness is modulated by the Euler-constant 'e'. e is at the root of exponential behavior. The vital by definition is about assertive and aggressive growth the symbol of which is 'e'. Hence vital-level (V) modulation (represented by 'mod') can be described by the following equation-segment:

$$V: Sig * mod\ (e)$$

If an organization exists at the mental level, it may be suggested that its uniqueness is modulated by the Gaussian Distribution 'G'. G summarizes rational behavior with a key direction followed by most, and directions more on the edge followed by outliers. Mental-level dynamics are arguably quite similar, and it can be suggested are best modeled by such a distribution (Salkind, 2007). Mental-level (M) modulation (mod) can hence be described by the following equation-segment:

$$M: Sig * mod\ (G)$$

The physical, the vital, and the mental levels are orientations in which patterns of perceiving, being, behaving are set in their ways. Each pattern has its purpose and its limitation and it can be argued that being able to learn from each orientation and yet being able to move beyond that, is the next logical step in any developmental model. The integral level hence, is about being able to leverage each of the patterns that naturally arise at the three preceding levels at will, and about further, being able to integrate these and arrive at new ways of perceiving and being.

Mathematically such behavior may be represented as being an integrative function ($\int x$) where 'x' is the ability to move between the patterns emanating from G, e, π, at will, represented by $G, \bar{e}, \pi$. Integral-level (I) modulation (mod) of uniqueness (*Sig*) can hence be represented by the following equation-segment:

$$I: Sig * mod\ (\int G, \bar{e}, \pi)$$

The condition of overcoming any fixed and limiting patterns is the prerequisite for the emergence of 'Force' or for entering into the force-level. At this level the uniqueness behind the particular development being considered can emerge in its purity and become a truly creative dynamic. This aspect of creativity that is in a sense not bound by circumstance may be represented by the constant 'c', the speed of light in a vacuum (Perkowitz, 2011) which is an upper limit of the layer that systems practically operate in. Force-level (F) modulation (mod) of uniqueness (*Sig*) can hence be represented by the following equation-segment:

$$F{:}\ Sig * mod\ (c)$$

Once the signature of an organization arises and continues to exercise itself in its purity, it achieves contextual-mastery (C) and is able to exercise itself as though the context it is acting in, that can vary in scale and complexity, were all of the same substance as itself. This equality may be represented by the integrative function '$\int = 1$'. The equation-segment that notates this contextual-level (C) modulation (mod) applied to organizational uniqueness (*Sig*) is hence:

$$C{:}\ Sig * mod\ (\int = 1)$$

Piecing all the equation-segments together the equation for the emergence of uniqueness (Sig_E), where 'X' can be any of the discussed modulations at the respective development-model levels (P, V, M, I, F, C), is hence summarized by Equation 2.5.1:

$$Sig_E = X \left| \begin{array}{c} C{:}Sig * mod\ (\int = 1) \\ F{:}Sig\ mod\ (c) \\ I{:}\ Sig\ mod\ (\int G, \bar{e}, \pi) \\ M{:}Sig * mod\ (G) \\ V{:}Sig * mod\ (e) \\ P{:}Sig * mod\ (\pi) \end{array} \right|$$

Eq 2.5.1: Emergence of Uniqueness

The power of virtual worlds in engaging people and promoting learning as described in "Why Virtual World Matter" (Thomas & Brown, 2009) can be seen as parallel to the emergence of uniqueness as described by Equation 2.5.1. As described by the authors there is a process of selectively leaving behind part of oneself (movement through the P, V, M levels) to recreate one's identity (movement through level I), and then engaging with others in a shared discourse and culture (movement through levels F, C), which becomes very meaningful for people.

Chapter 2.6: Inherent Dynamics of Any System

So far the inherent innovation that exists at the system level and summarized by the nature of a point has been considered. Further, how this deep fount of innovation is present everywhere and how sets that make more practical the range of creative forces available in each of the four components of a point have also been considered. These architectural forces further define the possibility inherent in any system. Leveraging these sets of forces an equation for the uniqueness of an organization, regardless of scale, was arrived at.

In some sense the precipitation of innovation from the barely perceptible nature of the ubiquitous point, to how this reveals a play of forces, to how organizations take their seed and grow from these forces, has been traced.

It is useful to now turn full-circle to return to the initial orientations that allowed so much to be suggested about the nature of innovation in the first place. It is useful to look deeper into the nature of the physical, the vital, the mental, and the integral, and to derive equations that in effect will provide further insight into the dynamics of innovation inherent in these orientations.

Hence, starting with the physical, an equation, Equation 2.6.1, is summarized as:

$$Physical = \begin{bmatrix} M_3 \rightarrow System_{Pr} \\ (\uparrow F \rightarrow I) \\ M_2 \rightarrow S_{Systemi_{Pr}} \\ (\uparrow Sig \rightarrow F) \\ M_1 \rightarrow Sig_P \\ (\uparrow > P_P) \\ U \rightarrow Physical_U \end{bmatrix} TC \rightarrow Physical_T$$

$$Where \begin{bmatrix} Physical_U \ni [inertia, lethargy, status\ quo, \ldots] \\ Physical_T \ni [adaptability, durability, strength, \ldots] \end{bmatrix}$$

Eq 2.6.1: Inherent Dynamism in Physical

Essentially this equation is laying out the conditions of moving from the untransformed or negative physical state represented by $Physical_U$ to the transformed or positive physical state represented by $Physical_T$.

The first matrix should be read from the bottom to the top:

$$\begin{bmatrix} M_3 \rightarrow System_{Pr} \\ (\uparrow F \rightarrow I) \\ M_2 \rightarrow S_{System_{Pr}} \\ (\uparrow Sig \rightarrow F) \\ M_1 \rightarrow Sig_P \\ (\uparrow > P_P) \\ U \rightarrow Physical_U \end{bmatrix}$$

Hence, at the bottom is the starting point '$U \rightarrow Physical_U$' which identifies the default or untransformed (U) level of the physical. The next row up, $(\uparrow > P_P)$, states that when the patterns of the untransformed physical (P_P) have been overcome (>), movement to the next level ($\uparrow$) is facilitated. Breaking through to the next level, $M_1 \rightarrow Sig_P$, allows its dynamics to become active. Hence, the signature or uniqueness of the physical (Sig_P) becomes active at meta-level 1 (M_1). As this signature becomes more like a Force ($Sig \rightarrow F$), the conditions for breakthrough ($\uparrow$) to the next level are achieved. This next level is referred to as meta-level 2 (M_2), and indicates that the architectural forces represented by the set of system-presence ($S_{System_{Pr}}$) have become more consciously active. When this Force becomes Integral $(F \rightarrow I)$ then the conditions for breakthrough ($\uparrow$) to the next level are achieved. The next level is notated as M_3 for meta-level 3, and the dynamics here indicate that the equation for system-presence becomes active. Becoming active basically means that the respective meta-level dynamic begins to act at the once 'untransformed' level (U) further modifying it. Modification or transformation began when M_1 became active. Transformation is accelerated when M_2 becomes active, and even further accelerated when M_3 becomes active. Note that the matrix is essentially semiotic in that adjacent levels exist in paired relationships and inter-related functionality (Usó-Doménech et al, 2016).

The notion of meta-layers is being explored by contemporary physicists and Erwin Laszlo in his book Self-Actualizing Cosmos (Laszlo, 2014) summarizes some of these developments: "Physicists describe the domain that underlies and embeds the particles, fields, and forces of the universe variously as quantum vacuum, physical spacetime, nuether, zero-point field, grand-unified field, cosmic plenum, or string-net liquid." Note that 'nuether' refers to a sub-quantum level of reality (Pearson, 1997). Laszlo goes on to describe a revolutionary discovery by Nika Arkani-Hamed of Princeton's Institute of Advanced Study, of a geometrical object, the amplituhedron (Arkani-Hamid et al, 2012), which is not in space-time but governs space-time so that it "appears that spatio-temporal phenomena are the consequence of geometrical relationships in a deeper dimension of physical reality". A deeper dimension of a physical layer suggests synonymity with a meta-layer.

The rate of the transformation can be better envisioned when considering action of the Transformation Circle, or TC. The TC can be thought of as 4 concentric circles, with M_3 at the center. M_3 is surrounded by M_2, which is surrounded by M_1. The outer circle is U. If TC is considered to be a clock, than at time 't = 0', the physical' can be thought of as being entirely in U. The clock starts ticking only when some initial patterns P_P are overcome $(>P_P)$. From this point on as time proceeds the conditions for breakthrough become riper, and a sinusoidal wave begins to integrate more of the concentric circles together. The sinusoid wave (sin) is itself modulated by an euler function, e^x, where 'x' is determined by the strength to overcome patterns ($\uparrow$) which will likely vary over time but will likely tend to be positive once the clock has started ticking because of the joy experienced with progressive movement. Being that the limit is the outer boundary of the concentric circles, there is further modulation by π until the 4 concentric circles have been integrated. TC, hence, may be represented by Equation 2.6.2:

$$TC \equiv (> P_P) \rightarrow \text{mod}\,(\sin, e^x, \pi)$$

Eq 2.6.2: Transformation Circle

Hence, the initial nature of the physical that may be characterized by the set comprising of elements such as, lethargy, acceptance of the status quo, amongst other such elements ($Physical_U \ni [inertia, lethargy, status\ quo, ...]$), transforms into a physical more characterized by elements such as adaptability, durability, strength, and so on ($Physical_T \ni [adaptability, durability, strength, ...$). This transformation represents the inherent innovation-dynamic within the Physical.

Similarly, the equation for the 'Vital', Equation 2.6.3, also shows the built-in transformation that represents the innovation-dynamic within the vital:

$$Vital = \begin{bmatrix} M_3 \rightarrow System_P \\ (\uparrow F \rightarrow I) \\ M_2 \rightarrow S_{System_P} \\ (\uparrow Sig \rightarrow F) \\ M_1 \rightarrow Sig_V \\ (\uparrow > P_V) \\ U \rightarrow Vital_U \end{bmatrix} TC \rightarrow Vital_T ,$$

$$Where \begin{bmatrix} Vital_U \ni [aggression, self\ centeredness, exploitation, ...] \\ Vital_T \ni [energy, support, adventure, enthusiasm, ...] \end{bmatrix}$$

Eq 2.6.3: Inherent Dynamism of Vital

The equation for the 'Mental', Equation 2.6.4, is similarly summarized as:

$$Mental = \begin{bmatrix} M_3 \rightarrow System_S \\ (\uparrow F \rightarrow I) \\ M_2 \rightarrow S_{System_S} \\ (\uparrow Sig \rightarrow F) \\ M_1 \rightarrow Sig_M \\ (\uparrow > P_M) \\ U \rightarrow Mental_U \end{bmatrix} TC \rightarrow Mental_T$$

$$Where \begin{bmatrix} Mental_U \ni [fixation, fundamentalism, fragmentation, ...] \\ Mental_T \ni [understanding, imagination, inspiration, ...] \end{bmatrix}$$

Eq 2.6.4: Inherent Dynamism of Mental

The equation for the 'Integral', Equation 2.6.5, is similarly summarized as:

$$Integral = \begin{bmatrix} M_3 \rightarrow System_N \\ (\uparrow F \rightarrow I) \\ M_2 \rightarrow S_{System_N} \\ (\uparrow Sig \rightarrow F) \\ M_1 \rightarrow Sig_I \\ (\uparrow > P_I) \\ U \rightarrow Integral_U \end{bmatrix} TC \rightarrow Integral_T$$

$$Where \begin{bmatrix} Integral_U \ni [possession, usurpation, hidden\ agendas, ...] \\ Integral_T \ni [appreciation, shift\ POV, MPV, synthesis, ...] \end{bmatrix}$$

Eq 2.6.5: Inherent Dynamism of Integral

The preceding equations can be generalized by Equation 2.6.6:

$$\boldsymbol{Innovation_{orientation-x}} = \begin{bmatrix} M_3 \rightarrow System_X \\ (\uparrow F \rightarrow I) \\ M_2 \rightarrow S_{System\,X} \\ (\uparrow Sig \rightarrow F) \\ M_1 \rightarrow Sig_x \\ (\uparrow > P_x) \\ U \rightarrow x_U \end{bmatrix} TC \rightarrow x_T \text{, where} \begin{bmatrix} x_U \ni [...] \\ x_T \ni [...] \end{bmatrix}$$

Eq 2.6.6: Generalized Equation of Innovation

In this generalized equation, $Innovation_{orientation-x}$, refers to the inherent innovation within a specific orientation. Orientation refers to the physical, the vital, the mental, or the integral.

Further, the notion of a core-matrix can be summarized by the following equation, Equation 2.6.7:

$$Core\_matrix = \begin{bmatrix} M_3 \rightarrow System_X \\ (\uparrow F \rightarrow I) \\ M_2 \rightarrow S_{System_X} \\ (\uparrow Sig \rightarrow F) \\ M_1 \rightarrow Sig_x \\ (\uparrow > P_x) \\ U \rightarrow x_U \end{bmatrix}$$

Eq 2.6.7: Core Matrix

One of the corollaries of the Generalized Equation of Innovation is that if the source of innovation is more influenced by a meta-level there will be simultaneity of innovation or emergence that becomes apparent at U. The higher the meta-level, the more likely that this simultaneity will be wider spread. In his book Where Good Ideas Come From: The Natural History of Innovation (Johnson, 2010) Johnson states: "A brilliant idea occurs to a scientist or inventor somewhere in the world, and he goes public with his remarkable finding, only to discover that three other minds had independently come up with the same idea in the past year." He refers to an essay "Are Inventions Inevitable" (Ogburn & Thomas, 1922) which uncovered 148 instances of similar yet independent innovation, most of them occurring within the same decade. Some examples include sunspots that were simultaneously discovered in 1611 by four scientists in four different countries, and the law of conservation of energy that was formulated separately four times in the late 1840s, amongst numerous other example.

Chapter 2.7: Qualified Determinism

Leveraging off the previous chapter a mathematical notion of qualified determinism is explored here. This

A new function, Dynamic Interaction (DI) that has a 'vertical' and a 'horizontal' component is introduced here. The vertical component is designated as DI_V and the horizontal component as DI_H. Several equations to capture the inherent dynamism that exists in each orientation or state have already been derived in Chapter 2.6. These included equations for the dynamism in the physical, the vital, the mental, and the integral. The derived equations propose a model to give insight into how innovation occurs by changing the fundamental states that an organization is subject to. Several scientists, such as (Prigogine, 1977) and others, are proposing that a system can bifurcate in unpredictable ways to create an emergent property that cannot be predicted. DI is going to propose that in fact there is a 'qualified determinism' as opposed to randomness that occurs.

This qualified determinism is the result of the relative strengths of the levels within the core-matrix identified in the derived equations and summarized by Equation 2.6.7. Hence, the generalized core-matrix, already is the following:

$$\begin{bmatrix} M_3 \rightarrow System_X \\ (\uparrow F \rightarrow I) \\ M_2 \rightarrow S_{System_X} \\ (\uparrow Sig \rightarrow F) \\ M_1 \rightarrow Sig_x \\ (\uparrow > P_{x)} \\ U \rightarrow x_U \end{bmatrix}$$

The application of the vertical component of the new function being proposed, DI_V, to this core matrix will yield the nature or 'strength' of the state (x) or orientation under consideration. If the untransformed or U layer is strongest, implying that the habitual patterns that keep an organization locked into its untransformed way of operation are still very active, then the nature of the output of DI_V, notated by x-state, will be x_U. If the habitual patterns have been overcome then the strength of the x-state increases since it is the dynamics of M_1 or Sig_x

that are now active. In this case the x-state will be Sig_x. If the unique 'signature' has become a 'force', then the conditions for activation of M_2 have been put in place and the x-state will be even higher, S_{System_X}. The architectural forces active in M_2 are by definition more powerful than Sig_x that is a derivation of a set of such architectural forces. If the 'force' so acting becomes impersonal so that an organizational ego-state is overcome, then the x-state will have the most strength and is characterized by $System_X$ active at M_3. Hence, DI_V applied to a core-matrix will yield the 'strength' in terms of the x-dynamic that is active. This is illustrated by the following equation, Equation 2.7.1 which can be considered to be a deductive proof in the context of this model:

$$DI_V \begin{bmatrix} M_3 \rightarrow System_X \\ (\uparrow F \rightarrow I) \\ M_2 \rightarrow S_{System_X} \\ (\uparrow Sig \rightarrow F) \\ M_1 \rightarrow Sig_x \\ (\uparrow > P_{x)} \\ U \rightarrow x_U \end{bmatrix} => $$

$$\underset{\sim}{x}\text{-state} \in \left(x_U, Sig_x, S_{System_X}, System_X\right)$$

$$Where: Strength(System_X) > Strength(S_{System_X}) > Strength(Sig_x) > Strength(x_U)$$

Eq 2.7.1: Illustrating Action of Dynamic Interaction – vertical component

What is to be noted here is that while the action of DI_V yields a relative strength and therefore a 'single' value for the core- or x-matrix under consideration yet each x-matrix in itself could have an infinite number of possibilities. This should be clear in looking at how x_U, Sig_x, S_{System_X}, and $System_X$, were initially defined.

Hence, taking the example where x = physical:

$Physical_U \ni [inertia, lethargy, status\ quo, ...]$

As can be seen $Physical_U$, defined in Chapter 2.6, is already an infinite set with qualities similar to the ones already specified.

Similarly, Sig_P, defined in Chapter 2.4, also has an infinite variation:

$$Sig_P = Xa + \overline{Yb_{0-n}} \text{ where } \begin{bmatrix} X \in [S_{System_{Pr}}] \\ Y \in [S_{System_{Pr}}, S_{System_P}, S_{System_K}, S_{System_N}] \\ a, b \text{ are integers}; a > b \end{bmatrix}$$

$S_{System_{Pr}}$, defined in Chapter 2.3, is also an infinite set with forces of the nature specified in the following equation:

$$S_{System_{Pr}} \ni [Service, Perfection, Diligence, Perseverance, ...]$$

And recall that in Chapter 2.2, $System_{Pr}$ has been defined as:

$$System_{Pr} \equiv TS_{0 \to N} \begin{bmatrix} \begin{matrix} Presence \\ \downarrow \\ Opportunity \end{matrix} & \begin{bmatrix} P_L & \to & V_L \\ V_L & \to & M_L \end{bmatrix} & \& \ Container_{\begin{bmatrix} System_P \\ System_K \\ System_N \end{bmatrix}} \end{bmatrix}$$

So in essence DI_V is really giving us a summary assessment of the 'level' of the x-matrix under consideration with all its infinite potentiality. An example will follow shortly.

The other component of DI, as suggested earlier in this chapter, is the horizontal component, DI_H. Just as DI_V yields a summary assessment of the level that an x-matrix is operating at, similarly DI_H yields a summary assessment of the direction that a system or organization under consideration is going to continue its development in considering the physical, the vital, the mental, and the integral orientations to be the choices.

Assuming that any organization or system is inherently unique, as this mathematical model proposes, and assuming that the infinite sets of x_U

and S_{System_X} applied across the physical, vital, mental, and integral orientations respectively will account for any state that an organization can experience, then at a certain point in time any organization under consideration is going to have a direction-bias in one of the possible physical, vital, mental, or integral directions. Hence, DI_H will yield the summary direction that is going to lead an organization into its future given the current states active in it.

This summary direction is going to be yielded by considering the relative strengths of the separate core x-matrices – the physical, the vital, the mental, the integral - under consideration. The assumption is that there will be one core-matrix that will be stronger than the others.

Hence, as an example, first applying DI_V across all four x-matrices may, for example, yield the following results, with the strongest level within each x-matrix highlighted and bolded:

$$\begin{bmatrix} System_{Pr} \\ \mathbf{S_{System_{Pr}}} \\ Sig_P \\ Physical_U \end{bmatrix} \begin{bmatrix} System_P \\ S_{System_P} \\ Sig_V \\ \mathbf{Vital_U} \end{bmatrix} \begin{bmatrix} System_K \\ S_{System_K} \\ Sig_M \\ \mathbf{Mental_U} \end{bmatrix} \begin{bmatrix} System_N \\ S_{System_N} \\ \mathbf{Sig_I} \\ Integral_U \end{bmatrix}$$

Since by definition the strength of $System_x$ is greater than S_{System_x}, which is greater than Sig_x, which is greater than x_U, applying DI_H, as in Equation 2.7.2, across these x-matrices, as in the example following it will then yield the strongest direction, which in this example is the Physical:

$$\mathbf{DI_H}\left(\begin{bmatrix} System_{Pr} \\ S_{System_{Pr}} \\ Sig_P \\ Physical_U \end{bmatrix} \begin{bmatrix} System_P \\ S_{System_P} \\ Sig_V \\ Vital_U \end{bmatrix} \begin{bmatrix} System_K \\ S_{System_K} \\ Sig_M \\ Mental_U \end{bmatrix} \begin{bmatrix} System_N \\ S_{System_N} \\ Sig_I \\ Integral_U \end{bmatrix}\right) = \mathbf{Orientation_{Strongest}}$$

Eq 2.7.2: Illustrating Action of Dynamic Interaction - horizontal component

Example:

$$DI_H\left(\begin{bmatrix} System_{Pr} \\ S_{System_{Pr}} \\ Sig_P \\ Physical_U \end{bmatrix}\begin{bmatrix} System_P \\ S_{System_P} \\ Sig_V \\ Vital_U \end{bmatrix}\begin{bmatrix} System_K \\ S_{System_K} \\ Sig_M \\ Mental_U \end{bmatrix}\begin{bmatrix} System_N \\ S_{System_N} \\ Sig_I \\ Integral_U \end{bmatrix}\right) = Physical$$

Hence, DI function will yield the following organizational direction, as in Equation 2.7.3, where 'x_matrix' is used interchangeably with 'orientation':

$$Org\_Dir = DI\left(\begin{matrix}\begin{bmatrix} M_3 \rightarrow System_{Pr} \\ (\uparrow F \rightarrow I) \\ M_2 \rightarrow S_{System_{Pr}} \\ (\uparrow Sig \rightarrow F) \\ M_1 \rightarrow Sig_P \\ (\uparrow > P_{P)} \\ U \rightarrow Physical_U \end{bmatrix} & \begin{bmatrix} M_3 \rightarrow System_P \\ (\uparrow F \rightarrow I) \\ M_2 \rightarrow S_{System_P} \\ (\uparrow Sig \rightarrow F) \\ M_1 \rightarrow Sig_V \\ (\uparrow > P_{V)} \\ U \rightarrow Vital_U \end{bmatrix} \\ \begin{bmatrix} M_3 \rightarrow System_S \\ (\uparrow F \rightarrow I) \\ M_2 \rightarrow S_{System_S} \\ (\uparrow Sig \rightarrow F) \\ M_1 \rightarrow Sig_M \\ (\uparrow > P_{M)} \\ U \rightarrow Mental_U \end{bmatrix} & \begin{bmatrix} M_3 \rightarrow System_N \\ (\uparrow F \rightarrow I) \\ M_2 \rightarrow S_{System_N} \\ (\uparrow Sig \rightarrow F) \\ M_1 \rightarrow Sig_I \\ (\uparrow > P_{I)} \\ U \rightarrow Integral_U \end{bmatrix}\end{matrix}\right) \rightarrow$$

$$x\_matrix_{strongest}\ @\ level_{strongest}$$

Eq 2.7.3: Organizational Direction

Generalizing, as in Equation 2.7.4, where Org_Dir is organizational direction:

$$Org\_Dir = DI\left(\begin{bmatrix} M_3 \rightarrow System_x \\ (\uparrow F \rightarrow I) \\ M_2 \rightarrow S_{System_x} \\ (\uparrow Sig \rightarrow F) \\ M_1 \rightarrow Sig_x \\ (\uparrow > P_P) \\ U \rightarrow x_U \end{bmatrix}_{x=p,v,m,i}\right) \rightarrow$$

$$x\_matrix_{strongest} \; @ \; level_{strongest}$$

Eq 2.7.4: Generalized Equation for Organizational Direction

Hence, this mathematical model is suggesting that any situation, rather than having a random outcome, has a 'qualified deterministic' outcome. In the introduction to his book "Where is Science Going?" (Planck, 1933), James Murphy points out that the reason Planck spent so much of his time giving lectures on causation was because of the trend of physicists at the time, which has continued to the modern day, to overthrowing the principle of causation following the development of quantum theory, which he felt was misplaced. "Planck would claim", he wrote, "and so would Einstein, that it is not the principle of causation itself which has broken down in modern physics, but rather the traditional formulation of it." Murphy also quotes James Jeans (Jeans, 1932) to suggest the issue associated with causation and determinism: "Einstein showed in 1917 that the theory founded by Planck appeared, at first sight at least, to entail consequences far more revolutionary than mere discontinuity", and here he is referring to the finding that radiant energy is not emitted in a continuous flow, but in integral quantities, or quanta, which can be expressed in integral numbers. Continuing: "It appeared to dethrone the law of causation from the position it had therefore held as guiding the course of the natural world. The old science had confidently proclaimed that nature could follow only one road, the road which was mapped out from the beginning of time to its end by the continuous chain of cause and effect; state A was inevitably succeeded by state B. So far the new science has only been able to say that state A may be followed by state B or C or D or by innumerable other states. It can, it is true, say that B is more likely than C, C than D, and so on; it can even specify the relative probabilities of B, C, and D. But, just because it has to speak in terms of probabilities, it cannot speak with certainty which state will follow which; this is a matter which lies on the knees of the gods – whatever gods there may be."

While under the apparent dynamics at the quantum level there may appear to be randomness and a dethroning of the principle of causation, the notion of a multiplicity of levels, each having its impact on the strength of an orientation and further on the consequent direction from a multiplicity of possible orientations, is being suggested here as determining the direction of any system, while still allowing infinite variation in the details that may define its. Hence, the positions of Planck and Einstein are vindicated when considering Equation 11.4.

Further, assuming any system where multiple elements are active, connected, interdependent, and emergent, it may be possible to understand, through application of calculus, as to which level is the source for change.

Hence, where N may be source of change, the rate of change of N will resolve into one of P_U, V_U, M_U, I_U, P_T, V_T, M_T, or I_T. This may be summarized by Equation 2.7.5, where y is either U or T:

$$\frac{dN}{dt} \rightarrow \begin{bmatrix} P_U & P_T \\ V_U & V_T \\ M_U & M_T \\ I_U & I_T \end{bmatrix} \rightarrow x_y, \text{ where } y \in (U, T)$$

Eq 2.7.5: Establishing the Nature of the Change

If T, implying that the action of one of the meta-levels has caused transformation, then application of one of the following integrals will determine which level is the likely source for change.

Hence, for M_1 if the integral of $\frac{\partial(x_U \rightarrow x_T)}{\partial t}$ across a limited area 'a' in the vicinity of the change, is greater than some threshold value $Threshold_{Signature}$, then the signature dynamics are likely the source of change. This is summarized by Equation 2.7.6:

$$\int_0^a \frac{\partial(x_U \rightarrow x_T)}{\partial t} > Threshold_{Signature}$$

Eq 2.7.6: Signature Dynamics as the Source of Change

For M_2, if the integral of $\frac{\partial(x_U \rightarrow x_T)}{\partial t}$ across a larger area 'b' extending beyond the vicinity of the change, is greater than some threshold value $Threshold_{Architectural\ Forces}$, then the architectural forces are likely the source of change. This is summarized by Equation 2.7.7:

$$\int_0^b \frac{\partial(x_U \rightarrow x_T)}{\partial t} > Threshold_{ArchitecturalForces}$$

Eq 2.7.7: Architectural Forces as the Source of Change

For M_3,, if the double integral of $\frac{\partial(x_U \rightarrow x_T)}{\partial t}$ across the system specified by 'A', and across some time 't', is greater than some threshold value $Threshold_{System\ Property}$, then the system properties are likely the source of change. This is summarized by Equation 2.7.8:

$$\int_0^t \int_0^A \frac{\partial(x_U \rightarrow x_T)}{\partial t} > Threshold_{SystemProperty}$$

Eq 2.7.8: System Properties as the Source of Change

Chapter 2.8: Mathematical Operators

Based on the mathematics developed in the previous chapters of Section 2, twelve general 'mathematical operators' are arrived at deductively. These operators are the result of considering some of the dynamics of each of the levels and integration of the multiple levels of the overall mathematical model. These are non exhaustive, but rather are indicative of the nature of operators that may arise in such a mathematical model.

The following operators have to be considered in context to the Generalized Equation for Innovation, Equation 2.6.6, derived in Chapter 2.6. This equation suggests that any system has implicit in it the urge to transform the untransformed layer, U, by opening to the influence of the meta-layers, M_1, M_2, and M_3. In so doing the very sources of innovation are altered and the visible characteristics of systems are transformed by the action of these sources of innovation. Several sets have already been suggested that explore these sources of innovation - $S_{System_{Pr}}$, S_{System_P}, S_{System_K}, and S_{System_N}, and the resultant characteristics of systems - $Physical_T$, $Vital_T$, $Mental_T$, and $Integral_T$.

The following sections leverage various aspects of the already derived mathematical model to frame sets of mathematical operators along each of the four bases of Light that components or elements of systems will be subject to in their journey toward transformation.

Presence-Based Mathematical Operators

Considering Light's dimension of Presence, Equation 2.8.1, Presence-Based Mathematical Operators, summarizes a set of representative presence-based mathematical operators:

$$Presence\_based_{Mathematical\_operators} \ni [Fullness, Equality, Uniqueness, \ldots]$$

Eq. 2.8.1: Presence-Based Mathematical Operators

Fullness refers to the possibility that every single point in any time-space continuum is informed by Light's four-fold fullness. Hence, in equation form it may be suggested that Fullness is the union (U) of $System_{Pr}$, $System_P$, $System_K$, and $System_N$ as in Equation 2.8.2, Fullness:

$$Fullness = \bigcup \begin{bmatrix} System_{Pr} \\ System_P \\ System_K \\ System_N \end{bmatrix}$$

Eq. 2.8.2: Fullness

In general for any two points 'A' and 'B' it can be suggested that the Fullness behind A is the same as the Fullness behind B, as in Equation 2.8.3, Equivalence of Fullness. This suggestion may also be arrived at by considering Einstein's General Theory of Relativity (Einstein, 1995) that states: "All bodies of reference K, K1 etc. are equivalent for the description of natural phenomena (formulation of the general laws of nature), whatever may be their state of motion." If we shrink these bodies of reference or coordinate systems to infinitesimal dimensions, thus approaching a 'point', this suggests that there is equivalence in that the general laws of nature are equally valid at any two points. Hence, Equation 2.8.3:

$$Fullness_A \equiv Fullness_B$$

Eq. 2.8.3: Equivalence of Fullness

Equality refers to the possibility that since every point in a time-space continuum is always some expression of Light's four-fold fullness, hence every point is equal to any other point. This implies that all expressions and developments share a fundamental equality with all other points. Depicting a point 'a' by $Point_A$ and a point 'b' by $Point_B$, then an equation, Equation 2.8.4, Equality, would be:

$$Equality\text{: } Point_A \equiv Point_B$$

Eq. 2.8.4: Equality

Uniqueness refers to the possibility that any point is fundamentally unique. An easy way to envision this is to see that any point in a time-space continuum is a result of a unique time-space intersection. Hence, two points within a time-space continuum, A and B, can always be envisioned as having unique time and space coordinates as in Figure 2.8.1:

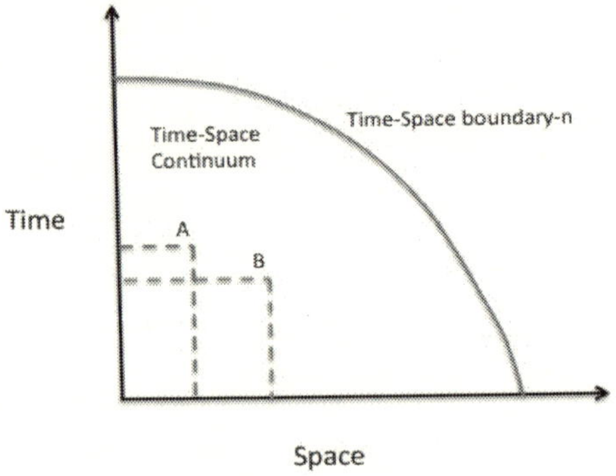

Figure 2.8.1: Uniqueness in Time-Space Continuum

But even for any organization, which can itself be considered a further development of a point or points through the action of time and space, the already proposed equation, (2.4.5) that governs organizational signatures may be restated as an equation for uniqueness, as in Equation 2.8.5:

$$Uniqueness\text{: } Sig = Xa + \overline{Yb_{0-n}} \quad where \begin{bmatrix} X \in [S_{System_{Pr}}, S_{System_P}, S_{System_K}, S_{System_N}] \\ Y \in [S_{System_{Pr}}, S_{System_P}, S_{System_K}, S_{System_N}] \\ a, b \text{ are integers}; a > b \end{bmatrix}$$

Eq. 2.8.5: Uniqueness

Power-Based Mathematical Operators

Power-based mathematical operators pertain to the fount of dynamism that will tend to determine an organization's practical action. Mathematical operators related to 'power' give insight into how an organization will tend to meet circumstance and other organizations.

Representative operators include Direction, Fractal, and Intersection and in contrast to Knowledge-based operators discussed in the next sub-section, tend to create the more visceral and immediate reactions to circumstance. Knowledge-based operators on the other hand tend to create action more in line with long-term or strategic plans. The integration between these different types of action and the notion of building such ambidexterity in organizations that combine incremental with perhaps more revolutionary shift is being recognized as increasingly important (Platzek, 2014). This power-based set can be summarized by Equation 2.8.6:

$$Power\_based_{Mathematical\_operators} \ni [Direction, Fractal, Intersection \ldots]$$

Eq. 2.8.6: Power-Based Mathematical Operators

Direction refers to the possibility that direction at any possible system bifurcation point is not random and not determined either. Rather it can be thought of as a qualified determinism, as introduced in Chapter 2.7, and is a function of applying DI_V and DI_H to the set of possibilities existing at a bifurcation point. As per the functioning of DI_V and DI_H, it is the strongest or most 'powerful' possibility that will tend to determine what will emerge as an organization meets circumstance and other organizations. Hence leveraging (2.7.4) creates Equation 2.8.7:

$$Direction\colon Org\_Dir = DI\left(\begin{bmatrix} M_3 \rightarrow System_X \\ (\uparrow F \rightarrow I) \\ M_2 \rightarrow S_{System_X} \\ (\uparrow Sig \rightarrow F) \\ M_1 \rightarrow Sig_x \\ (\uparrow > P_{P)} \\ U \rightarrow x_U \end{bmatrix}_{x=p,v,m,i}\right) \rightarrow x\_matrix_{strongest} @ level_{strongest}$$

Eq. 2.8.7: Direction

Fractal refers to the possibility that as organizational complexity increases, the base orientation, orientation-x, where x could be physical, vital, mental, or integral, of an average organization at some level of complexity 'n', will tend to determine the orientation of an organization at a level of complexity 'n+1'. Likewise the orientation of an organization at level of complexity 'n+1' will tend to determine the orientation of an organization at level of complexity 'n'. This is summarized by Equation 2.8.8:

$$\mathit{Fractal}\colon\ \boldsymbol{Orientation_x\ @\ Complexity_n \leftrightarrow Orientation_x\ @\ Complexity_{n+1}}$$

Eq. 2.8.8: Fractal

Organizational complexity refers to an order of magnitude change as in from a person to a team, or from a team to a business unit, and so on, for example. In Nature such fractal arrangements abound in the way the human body is constructed to the very structure of galaxies (Briggs, 1992). This notion has been suggested to exist in complex behavioral systems as well as described in some detail in books such as The Fractal Organization (Hoverstadt, 2008), The Fractal Organization (Malik, 2015), and The Misbehavior of Markets (Mandelbrot, 2006). This notion relates to 'power' in that it is the patterns at one level that will tend to determine the patterns at another level, often preempting what may be a more logical choice based on reason. This kind of behavior has been suggested as causing cyclic fluctuations in stock and other markets where greed and fear often trump more intelligent and rational choices (Frost, 2005). Greed rises until fear sets in. Fear rises until greed sets in.

Intersection occurs when two organizations shift orientations to the next successive level due to the shock of interaction. Hence if an organization is at a physical orientation, it may be shifted to a vital orientation when intersection occurs, as in Equation 2.8.9, Intersection, where the function 'Next Element' extracts the next element from the Set S comprising the elements (physical, vital, mental, integral). Examples of such phenomena abound where failure to make a shift results in shock of conflict repeating itself endlessly. Such a process with applicability at multiple levels of complexity has been captured by the series of books on crucial conversations (Patterson, 2011).

$$Intersection\colon\ Organization\ Orientation \rightarrow Next\ Element\ (S)$$

Eq. 2.8.9: Intersection

Knowledge-Based Operators

Knowledge-based mathematical operators have to do with how organizations tend to develop by existing or toward increasing knowledge over time. Representative knowledge properties discussed in this section include Alternative, Flowering, and Higher, which as suggested in the precious section are of a different nature than 'power' or dynamism-based properties that tend to determine an organization's visceral or immediate reaction to the market place. This knowledge-based set can be summarized by Equation 2.8.10:

$$Knowledge\_based_{Mathematical\_operators} \ni [Alternative, Flowering, Higher \ldots]$$

Eq. 2.8.10: Knowledge-Based Mathematical Operators

Alternative refers to an alternative narrative that an organization will tend to embed itself in. These alternative narratives relate to the physical, the vital, the mental and the integral orientations. These narratives can easily become fixed and can strongly influence the entire internal and external orientation of an organization. The book 'The Fractal Organization: The Future of Enterprise' (Malik, 2015) suggests a theory of such narratives with their consequent effect on practical action. The alternative narratives are best described using the generalized equation (2.6.6) modified as Equation 2.8.11. Hence:

$$Alternative: Innovation_{orientation-x} = \begin{bmatrix} M_3 \rightarrow System_X \\ (\uparrow F \rightarrow I) \\ M_2 \rightarrow S_{System X} \\ (\uparrow Sig \rightarrow F) \\ M_1 \rightarrow Sig_x \\ (\uparrow > P_x) \\ U \rightarrow x_U \end{bmatrix} TC \rightarrow x_T \text{, where } \begin{bmatrix} x_U \ni [\ldots] \\ x_T \ni [\ldots] \end{bmatrix}$$

Eq. 2.8.11: Alternative

Where, 'x' can be thought of as an element from the Set of Orientations:

$x \in (physical, vital, mental, integral)$

'X' can be thought of as an element from the Set of System-level architectural forces. Hence:

$X \in (Presence, Power, Knowledge, Nurturing)$

Flowering refers to the possibility that any time-space boundary-n, depicted as TS_n (as in Fig. 2.8.1) will have more potential or possibility associated with it than a time-space boundary-(n-1). Putting this into equation format yields Equation 2.8.12:

$$Flowering: Possibility_{TS_n} > Possibility_{TS_{n-1}}$$

Eq. 2.8.12: Alternative

Higher refers to the possibility that over time the direction will always tend to move to a higher meta-level. This can be summarized by using the notion of the core-matrix (2.6.7) yielding Equation 2.8.13:

$$Higher: Upward \begin{bmatrix} M_3 \rightarrow System_X \\ (\uparrow F \rightarrow I) \\ M_2 \rightarrow S_{System_X} \\ (\uparrow Sig \rightarrow F) \\ M_1 \rightarrow Sig_x \\ (\uparrow > P_{x)} \\ U \rightarrow x_U \end{bmatrix}$$

Eq. 2.8.13: Higher

Note also that while many organizations are practically at the untransformed level for a long time there is still movement within that level that can generally be depicted as a change from a predominantly physical-orientation to a more mental-orientation. In general this shift in orientation is implied, as more and more fixed patterns are overcome: $\uparrow > P_x$, where P_x refers to patterns along an orientation 'x' where x is an element from the set: (physical, vital, mental, integral).

Nurturing or Harmony-Based Mathematical Operators

Nurturing-based mathematical operators have to do with the nature of relationship within systems. These relationships are posited as being of a nurturing nature and emanate from the notion of 'nurturing' as an organizing class. Representative mathematical-operators include Remember, Linking, and Relate. This nurturing-based set can be summarized by Equation 2.8.14:

$$Nurturing\_based_{Mathematical\_operators} \ni [Remember, Linking, Relate \ldots]$$

Eq. 2.8.14: Nurturing-Based Mathematical Operators

Remember has to do with remembering that there is something in each organization that existed before the existence of any organization, and further, that there is something in each organization that exists in every other organization. This can be thought of going back to a time-space moment of zero, and subsequently of expanding into the Time-Space Continuum keeping that connection in mind. There is something in each organization that exists in every organization and highlights a special way to relate to the underlying system as in Figure 2.8.2:

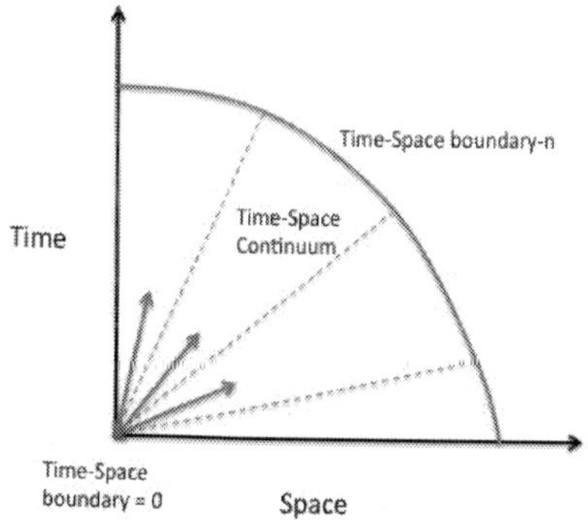

Figure 2.8.2: Remembrance in Time-Space Continuum

This may be depicted by Equation 2.8.15:

$$Remember: Ubiquity \begin{bmatrix} \leftarrow \\ TS = 0 \\ \rightarrow \\ TS > 0 \end{bmatrix}$$

Eq. 2.8.15: Remember

Where the condition of going back (←) before any organization existed, TS = 0, is invoked as all developments proceed (→), TS > 0, to create a sense of ubiquity. The sense of ubiquity is a remembrance. This notion may be akin to the concept that everything that is in the universe emanated from the Big Bang and that we are all recycled stardust (Swimme, 2001).

Link refers to the condition whereby in any time-space coordinate, irrespective of the level of untransformed reality (U), any present state can be consciously linked to the underlying ubiquitous system. This can be depicted by Equation 2.8.16, Link, where the conditions that usually need to be in place for a meta-layer to actively influence layer U disappear. There may be an attitude or receptiveness on the part of the element at U that allows such linking to take place. Hence:

$$Link: \begin{bmatrix} M_3 \rightarrow System_X \\ \uparrow\downarrow \\ M_2 \rightarrow S_{System_X} \\ \uparrow\downarrow \\ M_1 \rightarrow Sig_x \\ \uparrow\downarrow \\ U \rightarrow x_U \end{bmatrix}$$

Eq. 2.8.16: Link

Relate is a way to relate to the System so as to offer or surrender activities any kind of organization is involved in, to the System, and hence the Fullness or intelligence embedded in every point. This can be depicted by an Offer function, such that the first or relatively untransformed element in the function, depicted by x_U, is being offered to the second one, the union of the four-fold intelligence embedded in each point and depicted by the union function U[ ], as in Equation 2.8.17:

$$Relate:\ Offer\left(x_U, \bigcup \begin{bmatrix} System_{Pr} \\ System_P \\ System_K \\ System_N \end{bmatrix}\right)$$

Eq. 2.8.17: Relate

SECTION 3: THE MATHEMATICS OF QUANTIZATION

The foundational Light-Space-Time Emergence equation is derived and leveraged in subsequent quantization analyses. Basic interpretations of Schrodinger's wave equation and Heisenberg's uncertainty principle are also explored leading to a deeper look at space, time, energy, and gravity quantization and their subsequent impact on material reality:

- Equation for Light-Space-Time Emergence
- Speed of Light and Quanta
- Interpreting Schrodinger's Equation
- Interpreting Heisenberg's Uncertainty Principle
- A Deeper Look at Quantization of Space, Time, Matter, and Gravity
- Impact of Wavefunction and Quantization on Material Reality

Chapter 3.1: Equation for Light-Space-Time Emergence

Equation 2.6.6, the generalized equation for innovation, can be restated as an evolving form true for all time, as in Equation 3.1.1:

$$Innovation_{orientation-x} = \left(\begin{bmatrix} M_3 \rightarrow System_X \\ (\uparrow F \rightarrow I) \\ M_2 \rightarrow S_{System X} \\ (\uparrow Sig \rightarrow F) \\ M_1 \rightarrow Sig_x \\ (\uparrow > P_x) \\ U \rightarrow x_U \end{bmatrix} TC \rightarrow x_T \text{, where } \begin{bmatrix} x_U \ni [...] \\ x_T \ni [...] \end{bmatrix} \right)_{\langle x_U | x_T \rangle}$$

Eq 3.1.1: Evolving Form of Generalized Equation of Innovation

The added notation of $\langle x_U \mid x_T \rangle$ implies that the output of the previous iteration of the equation of innovation, x_T, where the subscript T implies relatively-transformed, now becomes the input, x_U, for the next iteration of the equation, where U implies relatively-untransformed. Hence through time there is greater and greater transformation that pushes experienced reality to greater and greater levels of functional-richness.

But further, given that quanta is proposed to be a doorway to deeper worlds of Light, that in fact allow aspects of those worlds or layers to become active at the surface layer U, the question is when have those aspects become active in manifest time. The following timeline based on generally accepted models of universal history (Particle Data Group, 2015) suggests when. Note too that the subsequent sections of emergence of the electromagnetic spectrum, matter, and life, will explore in far grater detail some of the statements made in the following timeline:

- At time, t $\leq$ 0 seconds, only M_3 is active. Recall that M_3 represents the four-fold reality present in every point-instant.

- At time, $0 \geq t > \infty$, M_2 the set of architectural forces continually gets added to, thereby increasing the size of the sets of forces.
- At time, $0 > t \geq 10^{-36}$ seconds, the equation of Innovation, $Innovation_{orientation-x}$, is such that M_1also becomes active, while U and TC are suppressed while present. The activation of M_1 begins to result in unique expressions or signatures of the set of architectural forces, and in this case in the reality of the essentially ubiquitous electromagnetic-spectrum (EM Spectrum) as a vehicle of the four-fold order that expressed itself in all that existed and in all that unfolded from that point in time on.
- At time, $t \sim 10^{-10}$ seconds, fundamental particles emerge as an essential material basis of the four architectural forces that frame all further development. As in the case of the EM Spectrum this implies the activity of M_1.
- At time, $t \sim 3 \times 10^5$ years light atoms emerge, and at time $t \sim 10^9$ years heavier atoms in the stars emerge. These also imply the continued activity of M_1.
- At time, $t \sim 13.8 \times 10^9$ years, the fourth clear expression of the same four-fold order as the bases of an even more complex organization, that of cellular life and all that is founded on it becomes clear. This time-point will be represented by the notation $t \sim E_{Cell}$, where 'E' stands for emergence. This too implies the activity of M_1. But further, TC becomes active here, with an implicit direction of operation from M_3 to U. Note that the sets of architectural forces specified by M_2 continue to increase the number of elements they comprise of as the complex interaction between the layers continues.
- At time $t > 13.8 \times 10^9$ years, human-beings, and more complex social organizations emerge. Here TC acts with an implicit direction of operation from U to M_3. This time-point will be represented by $t \sim E_{Human}$.

Based on the aforementioned description Equation 3.1.2 for Emergence true of any space-time scale may be generalized as the following:

$$Emergence_{space-time} = \left| \begin{matrix} \begin{bmatrix} M_3 \rightarrow System_X \\ (\uparrow F \rightarrow I) \\ M_2 \rightarrow S_{System_X} \\ (\uparrow Sig \rightarrow F) \\ M_1 \rightarrow Sig_x \\ (\uparrow > P_x) \\ U \rightarrow x_U \end{bmatrix}_{Space} \\ \left[U \rightarrow \begin{matrix} M_3 : -\infty \leq t \leq \infty \\ \downarrow \\ M_2 : 0 \geq t > \infty \\ \downarrow \\ M_1 : 0 > t > \infty \\ \downarrow \\ t \sim E_{Cell}; \text{TC: } M_3 \rightarrow U \\ t \sim E_{Human}; \text{TC: } U \rightarrow M_3 \end{matrix} \right]_{Time} \\ TC \rightarrow x_T \text{, where } \begin{bmatrix} x_U \ni [...] \\ x_T \ni [...] \end{bmatrix} \end{matrix} \right|_{\langle x_U | x_T \rangle}$$

Eq 3.1.2: Space-Time Emergence

An implication of this equation, brought out more explicitly through the elaboration of the 'Time' component, is that the layers U, M_1, M_2, and M_3 exist simultaneously. Adding the Light-Matrix derived in Chapter 2.1 enhances Equation 3.1.2 to the Light-Space-Time Emergence form as represented by Equation 3.1.3:

$$
Emergence_{light-space-time} = \left| \begin{array}{cc}
\left[\begin{array}{c}
C_{\infty}: [Pr, Po, K, H] \\
(\downarrow R_{C_K} = f(R_{C_{\infty}})) \\
C_K: [S_{Pr}, S_{Po}, S_K, S_H] \\
(\downarrow R_{C_N} = f(R_{C_K})) \\
C_N: f(S_{Pr}\ x\ S_{Po}\ x\ S_K\ x\ S_H) \\
(\downarrow R_{C_U} = f(R_{C_N})) \\
C_U: [P, V, M, C]
\end{array} \right]_{Light}
&
\left[\begin{array}{c}
M_3 \rightarrow System_X \\
(\uparrow F \rightarrow I) \\
M_2 \rightarrow S_{System_X} \\
(\uparrow Sig \rightarrow F) \\
M_1 \rightarrow Sig_x \\
(\uparrow > P_x) \\
U \rightarrow x_U
\end{array} \right]_{Space}
\\
\left[\begin{array}{c}
M_3 : -\infty \leq t \leq \infty \\
\downarrow \\
M_2 : 0 \geq \mathrm{t} > \infty \\
\downarrow \\
M_1 : 0 > \mathrm{t} > \infty \\
\downarrow \\
U \rightarrow \begin{array}{c} \mathrm{t} \sim E_{Cell};\ \mathrm{TC}: M_3 \rightarrow \mathrm{U} \\ \mathrm{t} \sim E_{Human};\ \mathrm{TC}: U \rightarrow M_3 \end{array}
\end{array} \right]_{Time}
&
TC \rightarrow x_T
\end{array} \right|_{\langle x_U | x_T \rangle}
$$

Eq 3.1.3: Light-Space-Time Emergence '

Chapter 3.2: Speed of Light and Quanta

As discussed conceptually in Section 1 and also mathematically in Chapter 2.1, since c is finite and therefore there is past, present, and future implied by it, this implies that at U a point has to become quanta. This is implicit in the notion of finiteness. Since light takes a finite amount of time to get from A to B, a "unit" of light will require a finite time to traverse that. Quanta at the subatomic level can be thought of as related to this finite time and distance for a unit of light to be expressed.

Planck's discovery that energy at the subatomic level requires a minimum threshold 'quanta' to express itself therefore makes sense. It is to be noted though that Planck's treatment of quanta was more as a mathematical convenience that allowed the derivation of an equation that explained the curve of radiation wave-lengths at varying temperatures of a heated black-body (Isaacson, 2008). Einstein though postulated quanta as a fundamental property of light itself, rather than as something that arose in the interaction of light with matter as Planck thought. Einstein's theory produced a law of the photoelectric effect where the energy of emitted electrons would depend on the frequency of light. Einstein received the Nobel Prize for this discovery (Isaacson, 2008).

Summarizing, if c is the upper limit of the layer U, then it makes sense that the lower limit h (Planck's constant) should be inversely proportional to c. Hence:

$$h \propto \frac{1}{c}$$

This relationship is substantiated by combining two well-known equations: the first is the electromagnetic equation connecting speed of light with wavelength and frequency, and the second is Einstein's photoelectric equation connecting energy with frequency of light:

(1) $C = \nu\lambda$
(2) $E = h\nu$

Yields:

$$h = \frac{E\lambda}{C}$$

About h, H.A. Lorentz the Dutch scientist has commented in The Science of Nature (Lorentz, 1925): "We have now advanced so far that this constant not only furnishes the basis for explaining the intensity of radiation and the wavelength for which it represents a maximum, but also for interpreting the quantitative relations existing in several other cases among the many physical quantities it determines. I shall mention a few only, namely the specific heat of solids, the photo-chemical effects of light, the orbits of electrons in the atom, the wavelengths of the lines of the spectrum, the frequency of the Roentgen rays which are produced by the impact of electrons of given velocity, the velocity with which gas molecules can rotate, and also the distances between the particles which make up a crystal. It is no exaggeration to say that in our picture of nature nowadays it is the quantum conditions that hold matter together and prevent it from completely losing its energy by radiation."

So just as c sets up the past-present-future experience and reality of U, h suggests that this experience will take place in shells of matter. In the absence of the limit h, as pointed out by Lorentz, only radiation, and no matter would exist. This reinforces the notion of the four-foldness implicit in the nature of light as already discussed in Sections 1 and 2.

The suggested variance in the speed of light by meta-layer may also throw some further light on the quantum realm. First, summarizing:

1. At U the speed of light in a vacuum, c_U, is finite at 186,000 miles/sec. This finiteness creates the reality and experience of past-present-future, and further a sense of fragmentation and separation. Further, assuming that c_U is a fundamental upper-limit at U, the inverse of it, $\frac{1}{C_U}$, must define some fundamental lower limit at U. This is indeed the case as Planck's constant, h, is proportional to this. 'h' allows for matter to be sustained, as it fundamentally limits the dispersion of energy as suggested by Lorentz.

2. At M_3, the speed of light, c_{M_3}, is suggested as being ∞ miles/sec. This allows a reality of 'oneness' and the possibility of a suggested four-fold intelligence existing in every point-instant of space-time as already suggested.
3. As also already suggested the quantum world, here designated by Q, because it is at boundary of U, accesses and interrelates with the meta-levels. As such, the speed of light, c_Q, will appear as a hybrid as in the following figure. Note though that it is really the speed of light at the native or resident layer that becomes active, and that this is simply being represented as c_Q for convenience:

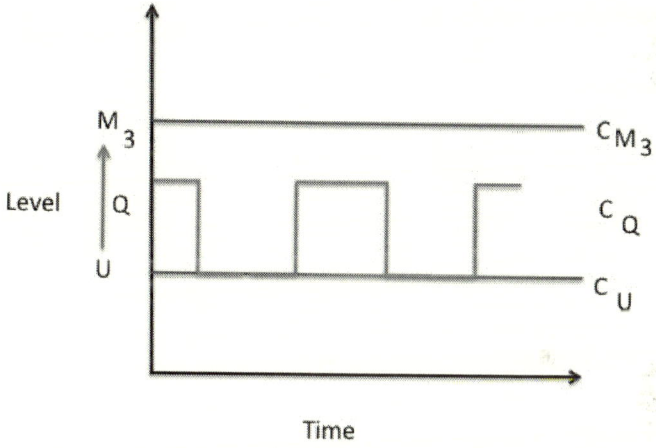

Figure 3.2.1 Speed of Light at Quantum Level

Note that research on the speed of light also indicates that it may go faster than c_U While the speeds suggested currently through experimental research, and summarized below, may be only incrementally higher than c_U the notion that c_U can be exceeded appears to be put in place:

1. The Heisenberg Uncertainty Principle (to be discussed subsequently in Chapter 3.4) already suggests that photons can travel at any speed, even exceeding c_U, for short periods.

2. Notion of different space-time realities, also known as meta-levels in this treatise, suggests that light can travel differently in a layer different from the four-dimensional space-time that apparently defines our observable world (Hawking, 1988)
3. In his book QED Feynman (Feynman, 1985) says "...there is also an amplitude for light to go faster (or slower) than the conventional speed of light. You found out in the last lecture that light doesn't go only in straight lines; now, you find out that it doesn't go only at the speed of light! It may surprise you that there is an amplitude for a photon to go at speeds faster or slower than the conventional speed, c." In research conducted at Humboldt University (Chown, 1990), Scharnhorst has made calculations using the theory of quantum electrodynamics to reveal the possible existence of "faster-than-light" photons. This is known as the Scharnhorst effect.
4. As reviewed in Chapter 2.1, Perkowitz makes the point that the theory of relativity does not disallow particles already moving at c or greater.

The point is that the reality at Q is going to be different than the reality at U. This should be apparent from considering the relation of c_X to the consequent reality as elaborated in Chapter 2.1. In Q the fundamental lower limit, h, which allows matter to sustain itself, is itself going to fluctuate. Hence, as X tends to M_3, c_X will tend to infinity, and h will become a fraction of itself. As it becomes a fraction of itself the quantization effect will be lowered, and matter will get dispersed more and more easily to in effect take on a wave-like, or field-like, or even block-like appearance as also suggested in Chapter 2.1.

Chapter 3.3: Interpreting Schrodinger's Equation

Schrodinger's equation, which seeks to model how a quantum state of a quantum system changes with time, or in other words seeks to model matter as a wave rather than as a particle (Stewart, 2012), is depicted in Equation 3.3.1:

$$i\frac{h}{2\pi}\frac{\partial}{\partial x}\psi = \hat{H}\psi$$

Eq 3.3.1: Schrodinger's Equation

ψ depicts a wave form and can be thought of as a probable cloud of possible states. $\hat{H}$ is the Hamiltonian operator which is a focusing function, and in its essence what the equation may be suggesting is that the way a wave form changes over time is equivalent to some expressible state of the possibilities inherent in the cloud of possible states.

But the cloud of possible states is another way of saying that behind the layer U form is represented in another way than at U. If the existence of the meta-levels, and in this case, of Siq_x at M_1, is considered possible, then it is far more reasonable to admit that form is configured by function and the very dynamics of what may appear to have been random, as some interpretations of quantum phenomena, may now appear to be far more logical. There is now more *context* to interpreting observation at the quantum level.

Interestingly Schrodinger himself had misgivings about the applicability of this equation that seemed to apply at the quantum level, to the macro-world (Stewart, 2012). To bring his misgivings to light he invented a thought experiment concerning a cat. This cat would be in a superposed state in a quantum black-box. A radioactive particle, a decaying-particle detector, and a flask of poison, were the other inhabitants of the black-box. At some point the particle will decay, be detected, and as in the thought-experiment at that point, triggered by the decaying particle, the poison in the flask would be released. The cat would then die. But in the meanwhile the cat would be in a superposed states of being both dead and alive. Only when the box was opened would the wave function collapse and a single definite state emerge.

Schrodinger was hoping to highlight the absurdity of the application of having a cat in both a dead and alive state at the macro-level. Instead physicists found this thought experiment to be sensible, even at the macro-level, and began to generalize the findings based on this. Hence, for example the idea of superposition at the physical level began to be thought of as real. It is interesting to note that in his lectures on Schrodinger's equation Feynman (Gottlieb, 2013) has stated: "Where did he get that [equation] from? Nowhere. It is not possible to derive it from anything you know. It came out of the mind of Schrödinger".

Considering Schrodinger's equation in the light of the discussion on Q in Chapter 3.2. 'i' is a complex number and suggests the interplay of two dimensions, one being real, and one being 'imaginary'. But the 'imaginary' dimension could be thought of as none other than the meta-levels implicit in the mathematical model in this treatise, and suggested to be real at Q. Further, $\frac{h}{2\pi}$ is in line with the suggestion also made in Chapter 3.2 that h will have to become a fraction of itself as c increases. Hence, the change in the wave function, $\frac{\partial}{\partial x}\psi$, is intimately related to i and $\frac{h}{2\pi}$, and perhaps more fully makes sense when considered in the context of i x $\frac{h}{2\pi}$ x $\frac{\partial}{\partial x}\psi$, which has to be the case when dealing with the integration of dynamics of multiple levels.

Further, the change in the wave function, $\frac{\partial}{\partial x}\psi$, is related to $\hat{H}\psi$, and suggests that there is some system "energy", represented by the Hamiltonian, $\hat{H}$, that when applied to the existing wave, ψ, will indicate how the wave will be expressed going forward.

But as discussed, at Q the dynamics of Sig_X become real, and in fact is a fundamental organizing principle for all organization at U, and starting at the dimension of h.

It may be the case that as the level of complexity of organization at the micro-level increases, as in from quantum, to atomic, to cellular, $\hat{H}$, itself complexifies as it houses further nuances or "superpositions" of Sig_X. Sig_X may be thought of as having more components as the complexity at such micro-levels of organization increases. Further, since it is being proposed here that all complexity or innovation is due to the integration of the dynamics of meta-levels with U, the representation of wave functions or ψ at the boundary world, Q, may be enhanced by consideration of Sig_X. Hence, $\hat{H}$ may be further qualified by notating $\hat{H}_Q$ or $\hat{H}_A$ or $\hat{H}_C$ for quantum, atomic, and cellular respectively, as per the following inequality in Equation 3.3.2:

$\hat{H}_Q < \hat{H}_A < \hat{H}_C$ (where: $\hat{H}_X \propto Energy_{Sig_X}$)

Eq 3.3.2: Hamiltonian Inequality

This inequality is specified by the number of components of Sig_X which it is assumed will have a direct effect on potential and kinetic energy of the wave-system (which are considered to be how $\hat{H}$ is measured) as in Equation 3.3.3:

$$Energy_{Sig_Q} < Energy_{Sig_A} < Energy_{Sig_C}$$

Eq 3.3.3: Quantum, Atomic, Cellular Energy Inequality

Similarly the following wave equations for quantum, atomic, and cellular respectively, as in Equations 3.3.4, 3.3.5, and 3.3.6 may be distinguished:

$$i\frac{h}{2\pi}\frac{\partial}{\partial x}\psi_Q = \hat{H}_Q\psi_Q$$

Eq 3.3.4: Wavefunction at Quantum Level

$$i\frac{h}{2\pi}\frac{\partial}{\partial x}\psi_A = \hat{H}_A\psi_A$$

Eq 3.3.5: Wavefunction at Atomic Level

$$i\frac{h}{2\pi}\frac{\partial}{\partial x}\psi_C = \hat{H}_C\psi_C$$

Eq 3.3.6: Wavefunction at Cellular Level

Chapter 3.4: Interpreting Heisenberg's Uncertainty Principle

In his book, The Little Book of String Theory, Princeton University's Gubser (Gubser, 2010) describes the effect on approaching absolute zero temperature on molecules. He takes the example of water molecules and relates that one cannot make the water molecules colder than absolute zero, -273.15 Celsius, because there is no more thermal energy to suck out at that temperature. However, quantum uncertainty, the phenomenon which relates the momentum and location of electrons in atoms necessitates that the water molecules will still vibrate. Gubser suggests this by considering Heisenberg's uncertainty relation, reproduced in Equation 3.4.1:

$$\Delta p \, X \, \Delta x \geq \frac{h}{4\pi}$$

Eq 3.4.1: Heisenberg's Uncertainty Relation

In Equation 3.4.1, Δp is the uncertainty in a particle's momentum, Δx is the uncertainty in the particle's location, and h is the Planck's constant. In frozen water crystals it is precisely known where the water molecules are, and therefore Δx is fairly small. This means that Δp has to be considerably larger, and therefore that the water molecules are still vibrating even though they are at absolute zero. This innate vibration, known as 'quantum zero-point' energy, expresses the phenomenon of quantum fluctuations.

The Planck's constant order of magnitude (10^{-34}) though, suggests the boundary between U and $M_1$and the quantum fluctuations, the uncertainty relation, and the quantum zero-point energy could be an expression of the essential Signature function, Sig_x, that is posited as a key formative force behind organization at U. In this interpretation the thermal energy describes the essential energy at U, while the uncertainty relation may suggest the phenomenon of innovation or function-precipitation "physically" linking M_1 and U. In this case it may be suggested that integration of meta-levels with the surface level, I_U^M, is indicated by the uncertainty relation, as in Equation 3.4.2:

$$I_U^M \rightarrow \Delta p \; X \; \Delta x \geq \frac{h}{4\pi}$$

Eq 3.4.2: Integration of Levels (Leveraging Heisenberg's Uncertainty Relation)

But further, it may also be suggested that the uncertainty principle itself is only valid at U, and that, because of the finiteness of c. This finiteness as already suggested implies h, which implies that if the position of a particle is going to be observed by shining light on it, the light has to have at least a quantum of energy. But to determine the position of a particle accurately light of a shorter wavelength would have to be used (Hawking, 1988) which would have to have a minimum amount of energy, which in turn would interfere with the velocity and hence momentum of the particle. The uncertainty in measuring the momentum could therefore be thought of as a consequence of the finiteness of the speed of light, c.

If c_U were to approach c_Q though, which as suggested in Chapter 3.2 could be anywhere between c and ∞ miles per second, the quantum would be smaller and the uncertainty in measuring position or momentum would be reduced. At c_{M_3} there would be no uncertainty since light would accurately tell both position and momentum definitively.

Hence, the uncertainty principle may be further qualified, as in Equations 3.4.3, 3.4.4, and 3.4.5:

$$@C_U\text{:} \; \Delta p \; X \; \Delta x \geq \frac{h}{4\pi}$$

Eq 3.4.3: Uncertainty Principle at U

$$@C_Q\text{:} \; \Delta p \; X \; \Delta x \rightarrow 0$$

Eq 3.4.4: Uncertainty Principle at Q

$$@C_{M_3}\text{:} \; \Delta p \; X \; \Delta x = 0$$

Eq 3.4.5: Uncertainty Principle at M3

The notion of position and momentum becoming finite at U also may imply that space, time, and quanta are emergent rather than absolute properties, as also suggested in Section 1. This is also the conclusion of Arkani-Hamed of the Institute of Advanced Studies in the following thought experiment (Wolchover, 2013):

'Locality says that particles interact at points in space-time. But suppose you want to inspect space-time very closely. Probing smaller and smaller distance scales requires ever higher energies, but at a certain scale, called the Planck length, the picture gets blurry: So much energy must be concentrated into such a small region that the energy collapses the region into a black hole, making it impossible to inspect. "There's no way of measuring space and time separations once they are smaller than the Planck length," said Arkani-Hamed. "So we imagine space-time is a continuous thing, but because it's impossible to talk sharply about that thing, then that suggests it must not be fundamental — it must be emergent."

Unitarity says the quantum mechanical probabilities of all possible outcomes of a particle interaction must sum to one. To prove it, one would have to observe the same interaction over and over and count the frequencies of the different outcomes. Doing this to perfect accuracy would require an infinite number of observations using an infinitely large measuring apparatus, but the latter would again cause gravitational collapse into a black hole. In finite regions of the universe unitarity can therefore only be approximately known.'

Chapter 3.5: A Deeper Look at Quantization of Space, Time, Matter, and Gravity

The Light-Space-Time Emergence equation (3.1.3) derived in Chapter 3.1, and reproduced here for convenience, offers some insight into the process of quantization. Reproducing the equation:

$$
Emergence_{light-space-time} = \left|\begin{array}{cc}
\left|\left[\begin{array}{c}
C_{\infty}: [Pr, Po, K, H] \\
(\downarrow R_{C_K} = f(R_{C_{\infty}})) \\
C_K: [S_{Pr}, S_{Po}, S_K, S_H] \\
(\downarrow R_{C_N} = f(R_{C_K})) \\
C_N: f(S_{Pr}\; x\; S_{Po}\; x\; S_K\; x\; S_H) \\
(\downarrow R_{C_U} = f(R_{C_N})) \\
C_U: [P, V, M, C]
\end{array}\right]_{Light}\right. &
\left.\left[\begin{array}{c}
M_3 \rightarrow System_X \\
(\uparrow F \rightarrow I) \\
M_2 \rightarrow S_{System_X} \\
(\uparrow Sig \rightarrow F) \\
M_1 \rightarrow Sig_x \\
(\uparrow > P_x) \\
U \rightarrow x_U
\end{array}\right]_{Space}\right| \\
\left[\begin{array}{c}
M_3 : -\infty \leq t \leq \infty \\
\downarrow \\
M_2 : 0 \geq t > \infty \\
\downarrow \\
M_1 : 0 > t > \infty \\
\downarrow \\
U \rightarrow \begin{array}{c} t \sim E_{Cell}; \text{TC: } M_3 \rightarrow U \\ t \sim E_{Human}; \text{TC: } U \rightarrow M_3 \end{array}
\end{array}\right]_{Time} & TC \rightarrow x_T
\end{array}\right|_{\langle x_U | x_T \rangle}
$$

The Light-Matrix suggests that there are particular kinds of quantization that could occur. Along the vertical realm these can be thought of as inter-relating one layer of the matrix with the previous layers. Hence the kind of quantization that is relevant to the physical layer, U, would relate c_U with c_x where x $\in$ (N, K, ∞), one of the possible speeds of c as per this mathematical treatise. This can be represented by h_U, where h stands for Planck's constant. Note that in this model there are several fundamental quantization possible that may inter-relate a specific light-layer with others above it. For the sake of simplicity it will be assumed that h_U is the quanta experienced in the inter-relation between U and the collective-set of light-layers behind or above U.

In looking at the Light-Matrix it is also clear that each subsequent layer below the top layer, describing the reality set up by an infinite speed, is emergent. Hence space, time, matter, and gravity, that arise when Light slows down, can be thought of as emergent phenomena. The emergence itself can be thought of as a function of Light-Matrix. At layer U, and as discussed in Chapter 1.2, Space is related to Light's property of Knowledge, Time is related to Light's property of Power, Matter or Energy is related to Light's property of Presence, and Gravity is related to Light's Property of Harmony or Nurturing. Each of these will emerge in a particular way and it is possible that there are multiple h_U's.

Hence, the Planck's constant for matter or energy, which we are already familiar with from past discoveries in science, can be depicted as h_{UPr}, since it is related to Presence (Pr). But similarly, quantization for space, time, and gravity could potentially be governed by other similar constants, referred to as h_{UK}, h_{UP}, and h_{UH}, and related to Knowledge (K), Power (P), and Harmony (H), respectively. Since the relationship between these constants, in absolute terms, is uncertain, this can be represented by the equality-inequality as depicted by Equation 3.5.1:

$$h_{UPr} \leq\geq h_{UK} \leq\geq h_{UP} \leq\geq h_{UH}$$

Eq 3.5.1: Equality-Inequality Relationship Between Different 'Planck' Constants

Further, the "quantization-window", that is, the window that quanta provides to layers of light behind U, as it were, potentially allows the precipitation of, or inter-relation with, or creation of a cohesive and compelling meta-function or signature as modeled by the C_N/M_1 layer. This quantization-window is positioned as being key in allowing a phenomenon of "quantum-certainty" to occur and will be explored in greater detail in the subsequent chapters on the working of quantization using history as an example. The 'Time' matrix, in (3.1.3) reproduced above, indicates the general default direction under which such quantum-certainty can occur at Layer U. Generally at the pre-human level it may proceed more automatically by the dynamics of the system itself, as represented by the segment: $t \sim E_{Cell}; TC{:}\ M_3 \rightarrow U$. Beyond this level it will generally occur through the action of a cohesive will, as represented by the segment: $E_{Human}; TC{:}\ U \rightarrow M_3$. The opening of this quantization-window is modeled by the 'Space' matrix and will happen when habitual patterns are overcome so that 'will' or 'want' becomes cohesive.

The specific quantization that are occurring along the space, time, energy/matter, and gravity dimensions are modeled by the following equations, that are all based on the Sig_x equations derived in Chapter 2.4. As a reminder (2.4.5) is reproduced here for convenience:

$$Sig = Xa + Y\bar{b}_{0-n}$$

$$where: \begin{bmatrix} X \in [S_{System_{Pr}}, S_{System_{P}}, S_{System_{K}}, S_{System_{N}}] \\ Y \in [S_{System_{Pr}}, S_{System_{P}}, S_{System_{K}}, S_{System_{N}}] \\ a, b \text{ are integers}; a > b \end{bmatrix}$$

The Structure of Space

The structure of Space, which holds the seeds of knowledge of all that will emerge, hence, is modeled in the following way as specified by Equation 3.5.2:

$$Space_{quantization} = h_{UK}(Xa + Y\bar{b}_{0-n})$$

$$where: \begin{bmatrix} X \in [S_{System_{K}}] \\ Y \in [S_{System_{Pr}}, S_{System_{P}}, S_{System_{K}}, S_{System_{N}}] \\ a, b \text{ are integers}; a > b \end{bmatrix}$$

Eq 3.5.2: Space Quantization

In this model 'space' as an emergent property of Light, is structured by infinite seeds of knowledge. But because the emergence is taking place in a layer of reality generally itself structured by a finite speed of light, c, it has to be quantized. The quantization assures that the knowledge is not dissipated, but can accumulate to create seeds, and therefore, the structure of space itself. It is assumed that there is some kind of 'Planck's constant' in effect, that is modeled as being specific to the way knowledge or space may be quantized – hence, h_{UK}.

The Structure of Time

The structure of Time, which holds the inevitability of the seeds or knowledge emerging in a phased maturity, hence, is modeled in the following way as specified by Equation 3.5.3:

$$Time_{quantization} = h_{UP}(Xa + Y\bar{b}_{0-n})$$

$$where: \begin{bmatrix} X \in [S_{System_P}] \\ Y \in [S_{System_{Pr}}, S_{System_P}, S_{System_K}, S_{System_N}] \\ a, b \ are \ integers; a > b \end{bmatrix}$$

Eq 3.5.3: Time Quantization

In this model 'time' as an emergent property of Light, is structured by an inevitable process of maturity, which due to its inevitability, is related to power. But because the emergence is taking place in a layer of reality generally itself structured by a finite speed of light, c, it has to be quantized. The quantization assures that the power is not dissipated, but can accumulate to express phased maturity, and therefore, the structure of time itself. It is assumed that there is some kind of 'Planck's constant' in effect, that is modeled as being specific to the way power or time may be quantized – hence, h_{UP}.

Energy & Matter

Energy, which through a process of containment can result in Matter, hence, is modeled in the following way as specified by Equation 3.5.4:

$$Energy_{quantization} = h_{UPr}(Xa + Y\bar{b}_{0-n})$$

$$where: \begin{bmatrix} X \in [S_{System_{Pr}}] \\ Y \in [S_{System_{Pr}}, S_{System_P}, S_{System_K}, S_{System_N}] \\ a, b \ are \ integers; a > b \end{bmatrix}$$

Eq 3.5.4: Energy Quantization

In this model 'energy' as an emergent property of Light, results in the reality of matter. But because the emergence is taking place in a layer of reality generally itself structured by a finite speed of light, c, it has to be quantized. The quantization assures that the energy is not dissipated, but can accumulate to create matter. Planck's constant is referred to as - h_{UPr}.

Gravity

Gravity, which holds seemingly distinct objects in the layer of reality created by c together in a harmony, hence, is modeled in the following way as specified by Equation 3.5.5:

$$Gravity_{quantization} = h_{UH}(Xa + Y\bar{b}_{0-n})$$

$$where: \left[\begin{array}{c} X \in [S_{System_N}] \\ Y \in [S_{System_{Pr}}, S_{System_P}, S_{System_K}, S_{System_N}] \\ a, b \; are \; integers; a > b \end{array} \right]$$

Eq 3.5.5: Gravity Quantization

In this model 'gravity' as an emergent property of Light, results in a harmonious collectivity of seemingly independent objects. But because the emergence is taking place in a layer of reality generally itself structured by a finite speed of light, c, it has to be quantized. The quantization assures that the harmony is not dissipated, but can accumulate to express more and more complex collectivities on large-scale. It is assumed that there is some kind of 'Planck's constant' in effect, that is modeled as being specific to the way harmony or gravity may be quantized – hence, h_{UH}.

Chapter 3.6: Quantization Effect of Organizational Transformation on Material Reality

The Light-Space-Time Emergence equation (3.1.3) being iterative, can be used to model emergence as it proceeds from simpler four-fold to more complex four-fold manifestations. A detailed example of this will be elaborated in Section 8 after having explored emergence of matter and life in Sections 4, 5, 6, and 7. However, in keeping with the current quantum-analysis paradigm of using wavefunction to explore outcomes at a particular space and time, this chapter will develop two general approaches, summarized by equations, to understanding the possible quantization effect of organizational transformation on material reality. While the natural line of development given the mathematical treatise explored in this book is to proceed with such an understanding using (3.1.3), the wavefunction depiction will be elaborated to suggest how (3.1.3) can be leveraged to provide further insight into a probability-based quantum-analytical approach.

(3.1.3) is reproduced below for convenience and a more concise form of it, the Simplified Version of Light-Space-Time Emergence, as in Equation 3.6.1 shall be utilized through the rest of this chapter.

$$
Emergence_{light-space-time} = \left\| \begin{bmatrix} C_\infty: [Pr, Po, K, H] \\ (\downarrow R_{C_K} = f(R_{C_\infty})) \\ C_K: [S_{Pr}, S_{Po}, S_K, S_H] \\ (\downarrow R_{C_N} = f(R_{C_K})) \\ C_N: f(S_{Pr}\ x\ S_{Po}\ x\ S_K\ x\ S_H) \\ (\downarrow R_{C_U} = f(R_{C_N})) \\ C_U: [P, V, M, C] \end{bmatrix}_{Light} \begin{bmatrix} M_3 \rightarrow System_X \\ (\uparrow F \rightarrow I)\ | \\ M_2 \rightarrow S_{System X} \\ (\uparrow Sig \rightarrow F) \\ M_1 \rightarrow Sig_x \\ (\uparrow > P_x) \\ U \rightarrow x_U \end{bmatrix}_{Space} \begin{bmatrix} M_3 : -\infty \leq t \leq \infty \\ \downarrow \\ M_2 : 0 \geq t > \infty \\ \downarrow \\ M_1 : 0 > t > \infty \\ \downarrow \\ U \rightarrow \begin{matrix} t \sim E_{Cell}; \text{TC}: M_3 \rightarrow \text{U} \\ t \sim E_{Human}; \text{TC}: U \rightarrow M_3 \end{matrix} \end{bmatrix}_{Time} TC \rightarrow x_T \right\|_{\langle x_U | x_T \rangle}
$$

Simplifying each of the main matrices in (3.1.3) hence, yields Equation 3.6.1, the Simplified Version of the Light-Space-Time Emergence equation:

$$
Emergence_{light-space-time} = \left|[L][S][T]TC \rightarrow x_T\right|_{\langle x_U | x_T \rangle}
$$

Eq. 3.6.1: Simplified Version of Light-Space-Time Emergence

Quantization Effect on Material Reality Using Simplified Version of Light-Space-Time Emergence

As detailed in Chapter 2.6 on the inherent dynamics of any system, while relatively transformed organizations have opened to the active influence of meta-levels, relatively untransformed organizations remain under the influence of untransformed physical, vital, mental, and integral dynamics at U.

Equations 3.6.2 through 3.6.4 summarize three types of organizations using (3.6.1). These are the untransformed-organization with only untransformed bases active, the organization-in-transition with mixed bases, and the transformed organization with bases relatively transformed due to the active influence of meta-levels, respectively. Hence:

$$Untransformed - organization = \left| |[L][S][T]TC \rightarrow x_T|_{\langle x_U \mid x_T \rangle} \right|_U$$

Eq. 3.6.2: Untransformed-Organization

$$Organization - in - transition = \left| |[L][S][T]TC \rightarrow x_T|_{\langle x_U \mid x_T \rangle} \right|_{U\&M_x}$$

Eq. 3.6.3: Organization-in-Transition

$$Transformed - organization = \left| |[L][S][T]TC \rightarrow x_T|_{\langle x_U \mid x_T \rangle} \right|_{M_x}$$

Eq. 3.6.4: Transformed-Organization

Now the question is what is the level of quantization caused by an organization as specified by (3.6.2-4)? For only if quantization takes place will there be an effect on the material fabric. Equation 3.6.5, Quantization Effect of Organization on Material Fabric, summarizes how this may happen:

$$\textit{Impact on Materal Fabric} = \begin{bmatrix} |[L][S][T]TC \rightarrow x_T|_{\langle x_U \mid x_T \rangle} \\ \times \\ \left(|mod(Z_Q)|_{Y>U} \right) \\ \ni \\ Z_Q, Z \in \mathbb{U}\ (Space, Time, Energy, Gravity) \end{bmatrix}$$

Eq. 3.6.5: Quantization Effect of Organization on Material Fabric

Line 1 in the matrix is simply (3.6.1). To understand if quantization is going to occur, the organization resulting from the application of Line 1 is ($\times$) 'modulated' (mod) by the fourfold space-time-energy-gravity quantization ((Z_Q)) so long as Y>U, that is, the operative bases are relatively transformed and hence have an active influence greater than U. The fourfold quantization is specified by ($\ni Z_Q, Z \in U\ (Space, Time, Energy, Gravity)$) where ($\ni$) the quantization being applied (Z) is each (U) of the elements of the set (Space, Time, Energy, Gravity).

Quantization Effect on Material Reality Using Wavefunction Form of Light-Space-Time Emergence

In Chapter 3.3 we were introduced to Schrodinger's wavefunction equation (3.3.1), reproduced below for convenience, which seeks to model matter as a wave rather than as a particle:

$$i\frac{h}{2\pi}\frac{\partial}{\partial x}\psi = \hat{H}\psi$$

This wavefunction equation deals with dynamics behind the surface, and particularly of as many as infinite potential superposed states that then collapse into a material possibility in time and space at U. Such superposition is another way of saying that there is wholeness behind the surface. In (3.6.1) such wholeness is represented by L, S, T, the Light, Space, and Time matrices respectively.

As explained by Neil Turok in his book The Universe Within (Turok, 2012), Euler's formula reproduced below, can be used to model many naturally occurring phenomena because of its sinusoidal oscillation between narrow bounds as x increases. Reproducing Euler's formula:

$$e^{ix} = \cos x + i\sin x$$

A key characteristic of this equation is that the sum of the squares of the ordinary and complex parts, on the right side of the equation, is one. In quantum theory this ensures that the probabilities for all possible outcomes add up to one. Dealing in probabilities become important when considering as many as infinite superposed states.

Further, in modeling material reality a modified notation of the Schrodinger wavefunction as interpreted by Feynman is leveraged. This version features the integral sign, $\int$, meaning that all terms to the right of it have to be summed up for all space and time till the moment when the wavefunction is required to be known.

Hence, combining (3.6.1) with Feynman's interpretation of the Schrodinger wavefunction, with the Euler formula yields the following equations, 3.6.6-11 for organizations:

$$\psi_{untransformed\,-\,organization} = \left| \int e^{i \int \| [L][S][T] \|} \right|_U$$

Eq. 3.6.6: Wavefunction for Untransformed-Organization

Note that in (3.6.6) the formulation specifying iteration, $\langle x_U \mid x_T \rangle$, becomes redundant since iteration is implied by the integral across space and time, and hence is removed. Further, (3.6.6) suggests that so long as the basis of an organization is untransformed, specified by the U following the vertical-brackets, the outcome is going to remain an untransformed organization, specified by $\psi_{untransformed\,-\,organization}$.

Further, as specified by Equation 3.6.7, the Probability-View of Untransformed-Organization, the bases of such organizations is going to be either the untransformed physical (P_U), the untransformed vital (V_U), the untransformed mental (M_U), or the untransformed integral (I_U):

$$|\psi_{untransformed-organization}|^2 = P_U^2 + V_U^2 + M_U^2 + I_U^2 = 1$$

Eq. 3.6.7: Probability-View for Untransformed-Organization

As specified by (3.6.7) the probability that any of these bases will be leveraged in the creation of an organization adds up to one. Note that the physical, the vital, the mental, and integral were discussed in Chapter 2.6 on Inherent Dynamics of a System.

Equation 3.6.8, Wavefunction for Organization-in-Transition, suggests the mixed bases for organizations, as specified by dynamics of both the untransformed (U) and the meta-levels (M_x), which therefore results in organization-in-transition.

$$\psi_{organization-in-transition} = \left| \int e^{i\int |[L][S][T]|} \right|_{U\&M_x}$$

Eq. 3.6.8: Wavefunction for Organization-in-Transition

As specified by Equation 3.6.9, Probability-View for Organization-in-Transition, the probability that any of the untransformed (U) and transformed (T) bases will be leveraged in the creation of an organization adds up to one:

$$|\psi_{organization-in-transition}|^2 = P_U^2 + P_T^2 + V_U^2 + V_T^2 + M_U^2 + M_T^2 + I_U^2 + I_T^2 = 1$$

Eq. 3.6.9: Probability View for Organization-in-Transition

Equation 3.6.10, Wavefunction for Transformed-Organization, suggests some transformed bases for an organization, as specified by dynamics of the meta-levels (M_x), which therefore results in a transformed-organization.

$$\psi_{transformed-organization} = \left| \int e^{i\int |[L][S][T]|} \right|_{M_x}$$

Eq. 3.6.10: Wavefunction for Transformed-Organization

As specified by Equation 3.6.11, Probability-View for Transformed-Organization, the probability that the transformed (T) bases will be leveraged in the creation of matter adds up to one:

$$|\psi_{transformed-organization}|^2 = P_T^2 + V_T^2 + M_T^2 + I_T^2 = 1$$

Eq. 3.6.11: Probability View for Transformed-Organization

Now the question is how would the very fabric of material existence be impacted based on the type of organization in effect and what might an equation that captured that look like? Equation 3.6.12, Quantization Effect of Organization on Material Fabric (Wavefunction Form), is such an equation:

$$\psi_{impact-on-material-fabric} = \begin{bmatrix} \left(\left|\int e^{i\int\|[L][S][T]\|}\right|_{Y=(U,U\&M_x,M_x)}\right) \\ \times \\ \left(\left|mod(Z_Q)\right|_{Y>U}\right) \\ \ni \\ Z_Q, Z \in \mathbb{U}\ (Space, Time, Energy, Gravity) \end{bmatrix}$$

Eq. 3.6.12: Quantization Effect of Organization on Material Fabric (Wavefunction Form)

Taking each line in the equation separately: Line 1&2 from the top combines (3.6.6), (3.6.8), and (3.6.10) to essentially state that the wave-function for impact-on-fabric while governed by the Light-Space-Time matrices in (3.1.3) will vary based on the bases that are active, hence $Y = (U,U\&M_x,M_x)$, where Y equates to the family of bases that may be active: untransformed (U), a mix (U & M_x), or M_x. Since the primary part of Line 2 falls under the $\int$ it means that the operation is iterative since dynamics are summed up for space and time to that point where the wavefunction is sought.

A quantization impact on the material fabric of reality will only take effect, where 'taking effect' is specified by (mod) or being modulated by, if the bases of operation for the organization is more than just the untransformed (U) bases, hence Y>U. The impact on the material fabric is the product (x) of Line 2 & 4, and the specific quantization (Z_Q) that will occur are each ($\mathbb{U}$) of the members ($\in$) of the set $(Space, Time, Energy, Gravity)$ so that the very fabric of reality is altered through corresponding space, time, energy, and gravity quantization.

In effect the altering of the fabric due to quantization, as discussed in the previous chapter sets up a field where a unique function or signature can have a material impact. An example of this dynamic will be explored in greater detail in Section 10.

SECTION 4: ELECTROMAGNETIC SPECTRUM

The electromagnetic spectrum is a technical way to refer to light, essentially because amongst its range of properties it displays a tangible and simultaneous electric and magnetic or "electromagnetic" reality as well. So as light becomes more concrete to us or as the possibilities within it begin to emerge, one of the first forms it takes is as the electromagnetic spectrum. It must be the case that the previously surfaced properties of light – Presence, Power, Knowledge, and Harmony – emerge so as to define the very architecture of the electromagnetic spectrum.

Chapter 4.1 hence will explore the notion of Harmony so created in any layer of reality due essentially to a different constant speed of light. Chapter 4.2 on the emergence of Light's property of Knowledge will relate the range of archetypes existing in the electromagnetic spectrum. Chapter 4.3, Power, relates the gradient of electromagnetic energies to the emergence of Light's property of Power. Chapter 4.4 on Presence relates the possibility of different kinds of masses to the different energies in the electromagnetic spectrum. Chapter 4.5, Light's Emergence as Electromagnetic Spectrum, summarizes the emergence of the electromagnetic spectrum in terms of the underlying Light-Space-Time Emergence equation.

Chapter 4.1: Harmony

To begin with, we had already looked at how the speed of light sets up the nature of reality because of its speed. So for light traveling at c, past, present, future, and the notion of separation due to creation of islands of matter is the reality. It seems then that c architects the possibility of interaction in our system and can therefore be thought of a projection from the property of Harmony to create the basis for a matter-based harmony. So being, it can be suggested that the nature of the resultant interactions allows matter-based organizations, regardless of scale, to come into their own, to grow into their boundaries, and to form bonds based on the sense of being separated from other perceived organizations. This notion of forming bonds seems also to be related to the "magnetic" in electromagnetic.

In equation form, as in Equation 4.1.1 it would therefore be possible to specify the nature of reality so set up by the electromagnetic spectrum (EM_Spectrum) moving at c:

$$EM\_Spectrum_{Speed} = $$

$$Xa + \overline{Yb_{0-n}} \quad where \begin{bmatrix} X \in [S_{System_N}] \\ Y \in [S_{System_{P_T}}, S_{System_P}, S_{System_K}, S_{System_N}] \\ a, b \text{ are integers}; a > b \end{bmatrix}$$

Eq 4.1.1: Speed of EM Spectrum

The notion of 'connection', 'growing into one's own', 'form bonds', amongst other attributes of such a reality can be seen as elements of the four sets of architectural forces. Hence 4.1.1 suggests the mathematical equation that so defines the nature of reality due to the speed of c of the electromagnetic spectrum.

This also implies that as the speed of c changes, and as discussed in Chapter 2.1, the nature of relationship in such a reality will also change. Perhaps it is that there are several different relationships, or types of harmonies possible.

Leveraging (2.1.2) the previously defined inequality for c_X, the following inequality as in Equation 4.1.2, Inequality for Variable c, can be specified:

$$Inequality_C\text{: } c_U < c_{M_1} < c_{M_2} < c_{M_3}$$

Eq 4.1.2: Inequality for Variable c

It may be suggested that the nature of the 'layer' will change as c takes on different constant values. This is depicted in Equations 4.1.3 through 4.1.6:

$$[Nature(U)]_{c_U} = Xa + \overline{Yb_{0-n}} \;\; where \begin{bmatrix} X \in [S_{System_N}] \\ Y \in [S_{System_{P_r}}, S_{System_P}, S_{System_K}, S_{S\square stem_N}] \\ a, b \text{ are integers}; a > b \end{bmatrix}$$

Eq 4.1.3: Nature of U as a Function of c@U

$$[Nature(M_1)]_{c_{M_1}} = Xa + \overline{Yb_{0-n}} \;\; where \begin{bmatrix} X \in [S_{System_N}] \\ Y \in [S_{System_{P_r}}, S_{System_P}, S_{System_K}, S_{System_N}] \\ a, b \text{ are integers}; a > b \end{bmatrix}$$

Eq 4.1.4: Nature of M1 as a Function of c@M1

$$[Nature(M_2)]_{c_{M_2}} = Xa + \overline{Yb_{0-n}} \;\; where \begin{bmatrix} X \in [S_{System_N}] \\ Y \in [S_{System_{P_r}}, S_{System_P}, S_{System_K}, S_{System_N}] \\ a, b \text{ are integers}; a \not> b \end{bmatrix}$$

Eq 4.1.5: Nature of M2 as a Function of c@M2

$$[Nature(M_3)]_{C_{M_3}} = Xa + \overline{Yb_{0-n}} \quad where \begin{bmatrix} X \in [S_{System_N}] \\ Y \in [S_{System_{Pr}}, S_{System_P}, S_{System_K}, S_{System_N}] \\ a, b \; are \; integers; a > |b \end{bmatrix}$$

Eq 4.1.6: Nature of M3 as a Function of c@M3

In every one of the preceding cases what is to be noticed is that the primary element X derives from the architectural set of nurturing, S_{System_N}.

Chapter 4.2: Knowledge

The electromagnetic spectrum contains a range of waves embedded in it. These waves enable many different applications with practical utility apparent in every day life. So for example there is a region in the electromagnetic spectrum that we call radio waves, and others that we know as microwave, infrared visible light, ultraviolet, x-rays and gamma rays that each make possible many technologies that we use every single day. These ranges essentially code a range of technological-possibility, and therefore the wave-range within the electromagnetic spectrum can be thought of as encoding or of expressing some kind of precipitation or projection or emergence of the property of Knowledge. Or put another way, the property of Knowledge that is found in light, emerges as the wave-range implicit in the electromagnetic spectrum.

The EM spectrum itself, with its vast range of natures from gamma rays through visible light through radio waves, with its implicitness of time-space possibility as suggested by frequency (ν) and wavelength (λ), may be thought of as an arrangement of archetypes of what is possible in systems, and therefore is perhaps a precipitation of system-knowledge, as in Equation 4.2.1:

$$EM\_Spectrum_{Structure} = Xa + \overline{Yb_{0-n}} \quad where \begin{bmatrix} X \in [S_{System_K}] \\ Y \in [S_{System_{P_T}}, S_{System_P}, S_{System_K}, S_{System_N}] \\ a, b \text{ are integers}; a > b \end{bmatrix}$$

Eq 4.2.1: Structure of EM Spectrum

What this also implies is that the significance or intent of the different types of waves that exist can also be expressed by this general equation where the X and Y elements will vary. What precisely these elements are will need to be worked out. A few representative equations, Equations 4.2.2 through 4.2.5 follow:

$$Gamma\ Rays_{\ Intent} = Xa + \overline{Yb_{0-n}}\ \ where \begin{bmatrix} X \in [S_{System_K}] \\ Y \in [S_{System_{P_r}}, S_{System_P}, S_{System_K}, S_{System_N}] \\ a, b\ are\ integers; a > b \end{bmatrix}$$

Eq 4.2.2: Gamma Rays Intent

Knowing that some of the applications of Gamma Rays are in sterilizing and radiotherapy, these can be attributed as elements to the architectural sets. The totality of the use of can be thought of as the 'intent' of Gamma Rays. An understanding of the uses will allow a full set of the secondary elements to be mapped thus allowing (4.2.2) to be reverse-engineered.

Similarly each of the archetypes present in the electromagnetic spectrum can be mapped out. Sample mapping of intent include x-rays, infrared, and microwaves:

$$X - Rays_{Intent} = Xa + \overline{Yb_{0-n}}\ \ where \begin{bmatrix} X \in [S_{System_K}] \\ Y \in [S_{System_{P_r}}, S_{System_P}, S_{System_K}, S_{System_N}] \\ a, b\ are\ integers; a > b \end{bmatrix}$$

Eq 4.2.3: X-Rays Intent

$$Infrared_{Intent} = Xa + \overline{Yb_{0-n}}\ \ where \begin{bmatrix} X \in [S_{System_K}] \\ Y \in [S_{System_{P_r}}, S_{System_P}, S_{System_K}, S_{System_N}] \\ a, b\ are\ integers; a > b \end{bmatrix}$$

Eq 4.2.4: Infrared Intent

$$Microwaves_{\ Intent} = Xa + \overline{Yb_{0-n}}\ \ where \begin{bmatrix} X \in [S_{System_K}] \\ Y \in [S_{System_{P_r}}, S_{System_P}, S_{System_K}, S_{System_N}] \\ a, b\ are\ integers; a > b \end{bmatrix}$$

Eq 4.2.5: Microwaves Intent

Chapter 4.3: Power

The range of different wave-types or wavelengths implicit in the electromagnetic spectrum also moves with different frequencies. Since the speed of light is a constant, and it is known that speed is the product of frequency and wavelength, the greater the wavelength the lower will be the frequency of the wave-type. Conversely the less the wavelength the higher will be the frequency of the wave-type. And energy or power of a wave-type will depend on its frequency.

Hence, gamma rays that have a lower wavelength will have a higher frequency, and higher energy associated with it. Radio waves on the other hand that have a higher wavelength will have a lower frequency and therefore lower energy associated with it. So implicit in the electromagnetic spectrum is a gradient of energy. But it is also known that the penetration power is dependent on frequency. Therefore, the higher the frequency or energy, the higher will be the power of the wave-type. Another way to say this is that the property of Power in Light emerges as the energy-gradient in the electromagnetic spectrum. This Power aspect, by the way, seems to have been captured by the "electro" in electromagnetic.

The energy-gradient implicit in the EM spectrum suggests the power and energy with which knowledge moves and is perhaps a precipitation of system-power, as in Equation 4.3.1:

$$EM\_Spectrum_{Energy} =$$

$$Xa + \overline{Yb_{0-n}} \;\; where \left[\begin{array}{c} X \in [S_{System_P}] \\ Y \in [S_{System_{Pr}}, S_{System_P}, S_{System_K}, S_{System_N}] \\ a, b \; are \; integers; a > b \end{array}\right]$$

Eq 4.3.1: EM Spectrum Energy

This energy as suggested is directly proportional to the frequency, of which there is an infinite range as predicted by Maxwell's equations. Different frequencies have different penetration profiles (HyperPhysics, 2016) and it may be suggested that the nature of the energy is also a precipitation of a range of to be determined meta-functions as suggested in some sample Equations 4.3.2 through 4.3.6. Hence:

$$\textit{Gamma Rays}_{\ Nature\ of\ Energy} = Xa + \overline{Yb_{0-n}} \quad where \left[\begin{array}{c} X \in [S_{System_P}] \\ Y \in [S_{System_{P_T}}, S_{System_P}, S_{System_K}, S_{System_N}] \\ a, b\ are\ integers; a > b \end{array} \right]$$

Eq 4.3.2: Gamma Rays Nature of Energy

$$X - Rays_{Nature\ of\ Energy} = Xa + \overline{Yb_{0-n}} \quad where \left[\begin{array}{c} X \in [S_{System_P}] \\ Y \in [S_{System_{P_T}}, S_{System_P}, S_{System_K}, S_{System_N}] \\ a, b\ are\ integers; a > b \end{array} \right]$$

Eq 4.3.3: X-Rays Nature of Energy

$$Ultraviolet_{\ Nature\ of\ Energy} = Xa + \overline{Yb_{0-n}} \quad where \left[\begin{array}{c} X \in [S_{System_P}] \\ Y \in [S_{System_{P_T}}, S_{System_P}, S_{System_K}, S_{System_N}] \\ a, b\ are\ integers; a > b \end{array} \right]$$

Eq 4.3.4: Ultraviolet Nature of Energy

$$White\ Light_{Nature\ of\ Energy} = Xa + \overline{Yb_{0-n}} \quad where \left[\begin{array}{c} X \in [S_{System_P}] \\ Y \in [S_{System_{P_T}}, S_{System_P}, S_{System_K}, S_{System_N}] \\ a, b\ are\ integers; a > b \end{array} \right]$$

Eq 4.3.5: White Light Nature of Energy

$$AM\ Radio\ Waves_{Nature\ of\ Energy} = Xa + \overline{Yb_{0-n}} \quad where \left[\begin{array}{c} X \in [S_{System_P}] \\ Y \in [S_{Sys\square em_{P_T}}, S_{System_P}, S_{System_K}, S_{System_N}] \\ a, b\ are\ integers; a > b \end{array} \right]$$

Eq 4.3.6: AM Radio Waves Nature of Energy

Chapter 4.4: Presence

If there is a large range of frequencies implicit in the electromagnetic spectrum then there is also the possibility of different types of masses implicit in the electromagnetic spectrum. Frequency determines energy, and mass and energy are related through Einstein's famous MC-squared equation. So pushing a little further it is not just that mass and energy are related, but a different kind of frequency or wave-type potentially allows a different type of mass to emerge. So the possibility of different types of mass seems to be related to the property of Presence in Light. In other words the property of Presence emerges as the possibility of different types of masses as suggested by the range of mass-possibilities that can emerge from the electromagnetic spectrum.

Mass can be thought of as a container at U within which all possibility happens. In other words it can be thought of as a precipitation or emergence of system-presence as depicted in Equation 4.4.1:

$$EM\_Spectrum_{Mass\_Possibility} = Xa + \overline{Yb_{0-n}} \quad where \begin{bmatrix} X \in [S_{System_{Pr}}] \\ Y \in [S_{System_{Pr}}, S_{System_P}, S_{System_K}, S_{System_N}] \\ a, b \text{ are integers}; a > b \end{bmatrix}$$

Eq 4.4.1: EM Spectrum Mass Possibility

If the frequencies are infinite, then the possibility of the 'types' of masses or matter is also infinite. The mystery of 'Dark Matter' suggested by scientists to be 27% of our universe, as opposed to 5% of visible matter (NASA-darkmatter, 2016) may have some relation to this. Dark Matter will be further explored in Section 11.

Hypothetically the Equations 4.4.2 through 4.4.6 depict a range of mass possibilities:

$$Gamma\ Rays_{\ Mass\_Possibility} = Xa + \overline{Yb_{0-n}}\ \ where \begin{bmatrix} X \in [S_{System_{Pr}}] \\ Y \in [S_{System_{Pr}}, S_{System_{P}}, S_{System_{K}}, S_{System_{N}}] \\ a, b\ are\ integers; a > b \end{bmatrix}$$

Eq 4.4.2: Gamma Rays Mass Possibility

$$Ultraviolet_{\ Mass\_Possibility} = Xa + \overline{Yb_{0-n}}\ \ where \begin{bmatrix} X \in [S_{System_{Pr}}] \\ Y \in [S_{System_{Pr}}, S_{System_{P}}, S_{System_{K}}, S_{System_{N}}] \\ a, b\ are\ integers; a > b \end{bmatrix}$$

Eq 4.4.3: Ultraviolet Mass Possibility

$$Blue\ Light_{\ Mass\_Possibility} = Xa + \overline{Yb_{0-n}}\ \ where \begin{bmatrix} X \in [S_{System_{Pr}}] \\ Y \in [S_{System_{Pr}}, S_{System_{P}}, S_{System_{K}}, S_{System_{N}}] \\ a, b\ are\ integers; a > b \end{bmatrix}$$

Eq 4.4.4: Blue Light Mass Possibility

$$Microwaves_{\ Mass\_Possibility} = Xa + \overline{Yb_{0-n}}\ \ where \begin{bmatrix} X \in [S_{System_{Pr}}] \\ Y \in [S_{System_{Pr}}, S_{System_{P}}, S_{System_{K}}, S_{System_{N}}] \\ a, b\ are\ integers; a > b \end{bmatrix}$$

Eq 4.4.5: Microwaves Mass Possibility

$$FM\ Radio\ Waves_{Mass\_Possibility} = Xa + \overline{Yb_{0-n}}\ \ where \begin{bmatrix} X \in [S_{System_{Pr}}] \\ Y \in [S_{System_{Pr}}, S_{System_{P}}, S_{System_{K}}, S_{System_{N}}] \\ a, b\ are\ integers; a > b \end{bmatrix}$$

Eq 4.4.6: FM Radio Waves Mass Possibility

Chapter 4.5: Light's Emergence as Electromagnetic Spectrum

As discussed previously the Light-Space-Time Emergence equation (3.1.3) being iterative, can be used to model emergence as it proceeds from simpler four-fold to more complex four-fold manifestations. But further, as implied by (3.6.5), the Quantization Effect of Organization on Material Fabric equation, any process of organization has the possibility of altering the material fabric or fabric of existence so long as the bases involved are driven primarily by a meta-level.

Application of Light-Space-Time Emergence Equation

As suggested by (3.1.3), reproduced below for convenience, the architecture and details of the electromagnetic spectrum can be seen to be the result of the application of the Light, Space, and Time matrices as will be elaborated:

$$
Emergence_{light-space-time} = \left[\begin{array}{cc}
\left[\begin{array}{c} C_\infty: [Pr, Po, K, H] \\ (\downarrow R_{C_K} = f(R_{C_\infty})) \\ C_K: [S_{Pr}, S_{Po}, S_K, S_H] \\ (\downarrow R_{C_N} = f(R_{C_K})) \\ C_N: f(S_{Pr}\ x\ S_{Po}\ x\ S_K\ x\ S_H) \\ (\downarrow R_{C_U} = f(R_{C_N})) \\ C_U: [P, V, M, C] \end{array}\right]_{Light} &
\left[\begin{array}{c} M_3 \rightarrow System_X \\ (\uparrow F \rightarrow I) \\ M_2 \rightarrow S_{System_X} \\ (\uparrow Sig \rightarrow F) \\ M_1 \rightarrow Sig_x \\ (\uparrow > P_x) \\ U \rightarrow x_U \end{array}\right]_{Space} \\
\left[\begin{array}{c} M_3 : -\infty \leq t \leq \infty \\ \downarrow \\ M_2 : 0 \geq t > \infty \\ \downarrow \\ M_1 : 0 > t > \infty \\ \downarrow \\ U \rightarrow \begin{array}{c} t \sim E_{Cell}; TC: M_3 \rightarrow U \\ t \sim E_{Human}; TC: U \rightarrow M_3 \end{array} \end{array}\right]_{Time} & TC \rightarrow x_T
\end{array} \right]_{(x_U|x_T)}
$$

Starting with the Light-Matrix, the top left-hand matrix in (3.1.3), the first line from the top, C_∞:$[Pr, Po, K, H]$, specifies the fundamental architecture of the electromagnetic spectrum. As explored in Chapters 4.1 on Harmony, 4.2 on Knowledge, 4.3 on Power, and 4.4 on

Presence, each of these aspects are an emergence of the fundamental properties of Light at ∞.

Line 3 in the Light-Matrix, C_K: $[S_{Pr}, S_{Po}, S_K, S_H]$, elaborates the sets for Presence, Power, Knowledge, and Harmony, each containing multiple elements. For example, as explored in Chapter 4.1 on Harmony, the following elements: 'connection', 'growing into one's own', 'form bonds', for example, were suggested to exist in the Set of Harmony or Nurturing, and were accessed to contribute to the operative reality so set up by virtue of light traveling at a certain speed.

Specifically, Line 5, C_{N}: $f(S_{Pr}\ x\ S_{Po}\ x\ S_K\ x\ S_H)$, suggests that unique seeds are created from a combination of the elements from all four sets, with a particular element leading, that in effect creates the distinctness of the vast variety possible in the electromagnetic spectrum.

Line 6, $(\downarrow R_{C_U} = f(R_{C_N}))$, specifies quantization between the layer where the seeds are formed, and the physical layer we are familiar with, and as explored in Chapter 3.5 and 3.6, will result in Line 7, C_U: $[P,V,M,C]$, hence changing the material fabric of existence. The possibilities represented by Lines 1 through 5 hence concretize through the quantization represented by Line 6 to become the electromagnetic spectrum with its physical (related to Presence), vital (related to Power), mental (related to Knowledge), and connection (related to Harmony) aspects now existing in material reality typified by Light moving at c. Note that just as Line 6 represents a process of quantization relating the layer of reality created by Light traveling at c with the antecedent layers, so too Lines 2 and 4 as previously discussed, also represent quantization of a more subtle kind that ultimately plays a critical part in allowing the material fabric to express infinite diversity.

Typically it is the process as captured by the Space-Matrix that will determine if Line 6 is activated. Specifically patterns at the untransformed layer, U, will need to be overcome, as specified by the second-line from the bottom of the Space-Matrix: $(\uparrow > P_x)$. But as specified by the bottom-line of the Time-Matrix, reproduced below, it is only with the advent of the human-system that the automaticity of the action of meta-levels is reversed:

$$U \rightarrow \begin{matrix} t \sim E_{Cell};TC: M_3 \rightarrow U \\ t \sim E_{Human};TC: U \rightarrow M_3 \end{matrix}$$

Hence in the case of the electromagnetic spectrum system, which in this emergence is a pre-human system, the fact that patterns do not need to be overcome means that quantization happens automatically.

Application of Quantization Effect of Organization on Material Fabric Equation

Nonetheless, and given this context it is useful to review Equation 3.6.5, Quantization Effect of Organization on Material Fabric:

$$\textit{Impact on Materal Fabric} = \begin{bmatrix} |[L][S][T]TC \rightarrow x_T|_{(x_U|x_T)} \\ \times \\ \left(|mod(Z_Q)|_{Y>U}\right) \\ \ni \\ Z_Q, Z \in \mathbb{U}\ (Space, Time, Energy, Gravity) \end{bmatrix}$$

Line 1 in the matrix is simply (3.6.1) the Simplified Light-Space-Time Emergence equation. To understand if quantization is going to occur, the organization resulting from the application of Line 1 is, where 'is means 'multiplied' or $(\times)$ by a 'modulation', specified by (mod), of a fourfold space-time-energy-gravity quantization (Z_Q) kind, so long as Y>U, that is, the operative bases are relatively transformed and hence have an active influence greater than U. The fourfold quantization is specified by $(\ni Z_Q, Z \in U\ (Space, Time, Energy, Gravity))$ where $(\ni)$ the quantization being applied (Z) is each (U) of the elements of the set (Space, Time, Energy, Gravity).

But as just summarized in the Time-Matrix in (3.1.3) Y is by definition greater than U and hence quantization is automatic. In terms of the electromagnetic spectrum such quantization implies that wholeness becomes fully active through specific space, time, energy, and gravity quantization to create an holistic "ecosystem" with its own "electromagnetic spectrum logic" as it were. The wholeness has now precipitated into the material fabric and is available to be consciously and unconsciously tapped into.

Note that while the electromagnetic spectrum as one of the earlier material manifestations of Light exists pervasively, technologies that leverage its deeper or implicit aspects, such as the intent of various parts of the spectrum, will need the full operation of the Time-Matrix starting with the overcoming of patterns at U by human systems, in order to be fully leveraged. Such dynamics that in a sense 'evolve' the functioning of the electromagnetic spectrum will be reviewed in more detail in Section 10.

SECTION 5: EMERGENCE OF MATTER

So we find that the four underlying properties of Light that we call Harmony, Knowledge, Power, and Presence are of the essence of the speed with which the electromagnetic spectrum moves, the wave-range within the electromagnetic spectrum, the energy-gradient within the electromagnetic spectrum, and the mass-possibilities due to the electromagnetic spectrum, respectively. Further the description – electromagnetic – seems to have captured the Power-Harmony aspects implicit in light. In reality the electro-magnetic spectrum can likely be more completely described as wavearchetype-electro-magnetic-masspotential spectrum.

But further, we will find that layers of matter – quantum particles, which include bosons, and atoms – are also structured or emerge along the same property-lines or property-families of light.

Hence the following chapters, 5.1, 5.2, and 5.3, describe Light's emergence as quantum particles, bosons, and atoms respectively.

Chapter 5.1: Quantum particles

The electromagnetic spectrum or perhaps more accurately, the wavearchetype-electro-magnetic-masspotential spectrum, describes one of the first layers or translation of light into manifestation. This spectrum or field is an example of what stands behind visible matter and perhaps it is fair to say that all matter emerges from such spectrums.

As already explored the finite speed of such spectrums allows the build-up of energies or quanta, and this in turn may express itself as a series or as an array of quantum particles. The vast number of quantum particles that have so far been discovered has in turn yielded what is known as the Standard Model. This model is made up of what is called Quarks, Leptons, and Bosons, and a Higgs-Boson (Cottingham, 2007).

But if we look at these four fundamental quantum categories of particles from a property-based, or "functional" viewpoint a different chapter in the exploiration of light emerges.

Quarks

For instance, it can be seen first that the nucleus of an atom is made up of a combination of quarks. Specifically, a proton is composed of two "up" quarks and one "down" quark. Quarks have unusual names – up, down, charm, strange, top, bottom, with each subsequent pair belonging to a different "generation" of quarks A neutron is composed of two "down" quarks and one "up" quark. Protons and neutrons together make up the nucleus of an atom. But we also know that the number of protons in the nucleus specifies the Atomic Number of an atom. Atomic number in turn uniquely identifies the element from the periodic table. Hence, an atomic number of 47, for example, specifies that the element is Silver. In other words it can be suggested that the unique properties of an element, the knowledge of what it is and how it will behave in the universe, is related to the quark. It may be suggested that quarks, therefore, are associated with the precipitation or emergence of Light's property of Knowledge in the quantum world.

Hence, it could be that the signature for the family of quarks, as in Equation 5.1.1, is:

$$Sig_{quarks} = Xa + \overline{Yb_{0-n}} \quad where \begin{bmatrix} X \in [S_{System_K}] \\ Y \in [S_{System_{P_T}}, S_{System_P}, S_{System_K}, S_{System_N}] \\ a, b \text{ are integers}; a > b \end{bmatrix}$$

Eq 5.1.1: Generalized Signature of Quarks

As can be seen the primary element X is derived from the set of knowledge, S_{System_K}. Various elements, derived from the four sets would define the behavior of quarks and could be functions such as 'composite-arrangements', 'specifying attributes' amongst others, hence collectively describing quarks' way of being.

Leptons

When considering leptons it is useful to know that unlike quarks that only exist in composite arrangements with other quarks, leptons are solitary, point-like particles without internal structure (Arabatzis, 2006). The best-known lepton is the electron. So the electron may be considered as a surrogate for the lepton class. The electron appears to be associated with the flow of energy and power. Further they appear to be the adventurers easily leaving the atom they are a part of. They also form locks or bonds with other atoms through the force of attraction and repulsion. In some sense they seem to be a representation or precipitation or emergence of Light's property of Power.

The signature for the family of leptons, as in Equation 5.1.2, is:

$$Sig_{leptons} = Xa + \overline{Yb_{0-n}} \quad where \begin{bmatrix} X \in [S_{System_P}] \\ Y \in [S_{System_{P_T}}, S_{System_P}, S_{System_K}, S_{System_N}] \\ a, b \text{ are integers}; a > b \end{bmatrix}$$

Eq 5.1.2: Generalized Signature of Leptons

As can be seen the primary element X is derived from the set of power, S_{System_P}. Various elements, derived from the four sets would define the behavior of leptons and could be functions such as 'adventurer', 'solitary', 'create attraction', amongst others, hence collectively describing leptons' way of being.

Bosons

The bosons on the other hand are thought of as force-carriers. They are what allow all known matter particles to interact. The three fundamental bosons in this category are the photon, the W and Z bosons, and the gluon. The carrier particle of the electromagnetic force or spectrum is the photon. The carrier particle of the strong nuclear force that holds quarks together is the gluon. The carrier particle for the weak interactions, responsible for the decay of massive quarks and leptons into lighter quarks and leptons, are the W and Z bosons.

Bosons can be thought of as the precipitation of what creates relationship and harmony at the quantum level. Hence they can be thought of as the precipitation or emergence of Light's property of Harmony at the quantum level.

The signature for the family of gauge bosons, as in Equation 5.1.3, is:

$$Sig_{bosons} = Xa + \overline{Yb_{0-n}} \quad where \begin{bmatrix} X \in [S_{System_N}] \\ Y \in [S_{System_{Pr}}, S_{System_P}, S_{System_K}, S_{System_N}] \\ a, b \text{ are integers}; a > b \end{bmatrix}$$

Eq 5.1.3: Generalized Signature of Bosons

As can be seen the primary element X is derived from the set of nurturing, S_{System_N}. Various elements, derived from the four sets would define the behavior of bosons and could be functions such as 'creating interaction', 'holding together', amongst others, hence collectively describing bosons' way of being.

Equations in this form give a clue as to how to think about the fundamental particles. While the default approach is always to think about physical qualities, equations in the preceding format allow us to think about detailed functions or qualities associated with particles and give insight into how the process of emergence or complexification of four-fold functionality takes place.

Higgs-Boson

This leaves the other discovered fundamental particle the Higgs-Boson. In ordinary matter, most of the mass is contained in atoms, and the majority of the mass of an atom resides in the nucleus, made of protons and neutrons. Protons and neutrons are each made of three quarks. But it is the quarks that get their mass by interacting with what is known as the Higgs field (Olive, 2014). Hence the Higgs-Boson can be thought of as the mass-giver. In other words it is what gives presence to the quarks and it can be thought of as the precipitation or emergence of Light's property of Presence at the quantum level. Just as there are multiple particles in each of the other 'families' it is likely that there will be multiple particles in the Higgs-Boson family. Recent research at CERN indicates that the Higgs-Boson may have a cousin.

The signature for the Higgs-boson and any other similar particle, as in Equation 5.1.4, is:

$$Sig_{Higgs-boson} =$$

$$Xa + \overline{Yb_{0-n}} \quad where \begin{bmatrix} X \in [S_{System_{Pr}}] \\ Y \in [S_{System_{Pr}}, S_{System_{P}}, S_{System_{K}}, S_{System_{N}}] \\ a, b \; are \; integers; a > b \end{bmatrix}$$

Eq 5.1.4: Generalized Signature of Higgs-Boson

As can be seen the primary element X is derived from the set of presence, $S_{System_{Pr}}$. Various elements, derived from the four sets would define the behavior of Higgs-bosons and could be a function such as 'creating mass', amongst others, hence collectively describing Higgs-bosons' way of being.

Combining all the preceding particle equations it is possible to create a generalized particle equation, as in Equation 5.1.5:

$$Sig_{particle} = Xa + \overline{Yb_{0-n}} \quad where \begin{bmatrix} X \in [S_{system_{Pr}}, S_{system_{P}}, S_{system_{K}}, S_{system_{N}}] \\ Y \in [S_{system_{Pr}}, S_{system_{P}}, S_{system_{K}}, S_{system_{N}}] \\ a, b \ are \ integers; a > b \end{bmatrix}$$

Eq 5.1.5: Generalized Signature of Particle

So we see that again the underlying properties of light - Knowledge, Power, Harmony, and Presence - emerge as quarks, leptons, bosons, and Higgs-bosons respectively.

Light's Emergence as Quantum Particles

As discussed previously the Light-Space-Time Emergence equation (3.1.3) being iterative, can be used to model emergence as it proceeds from simpler four-fold to more complex four-fold manifestations. But further, as implied by (3.6.5), the Quantization Effect of Organization on Material Fabric equation, any process of organization has the possibility of altering the material fabric or fabric of existence so long as the bases involved are driven primarily by a meta-level.

As suggested by (3.1.3), reproduced below for convenience, the architecture and details of quantum particles can be seen to be the result of the application of the Light, Space, and Time matrices as will be elaborated:

$$
Emergence_{light-space-time} = \left| \begin{matrix} \begin{bmatrix} C_{\infty}: [Pr, Po, K, H] \\ (\downarrow R_{C_K} = f(R_{C_\infty})) \\ C_K: [S_{Pr}, S_{Po}, S_K, S_H] \\ (\downarrow R_{C_N} = f(R_{C_K})) \\ C_{N:} f(S_{Pr}\ x\ S_{Po}\ x\ S_K\ x\ S_H) \\ (\downarrow R_{C_U} = f(R_{C_N})) \\ C_U: [P, V, M, C] \end{bmatrix}_{Light} & \begin{bmatrix} M_3 \rightarrow System_X \\ (\uparrow F \rightarrow I) \\ M_2 \rightarrow S_{System X} \\ (\uparrow Sig \rightarrow F) \\ M_1 \rightarrow Sig_x \\ (\uparrow > P_x) \\ U \rightarrow x_U \end{bmatrix}_{Space} \\ \begin{bmatrix} & M_3 : -\infty \leq t \leq \infty \\ & \downarrow \\ & M_2 : 0 \geq \mathrm{t} > \infty \\ & \downarrow \\ & M_1 : 0 > \mathrm{t} > \infty \\ & \downarrow \\ U \rightarrow & \begin{matrix} \mathrm{t} \sim E_{Cell}; \mathrm{TC}: M_3 \rightarrow \mathrm{U} \\ \mathrm{t} \sim E_{Human}; \mathrm{TC}: U \rightarrow M_3 \end{matrix} \end{bmatrix}_{Time} & TC \rightarrow x_T \end{matrix} \right|_{\langle x_U | x_T \rangle}
$$

Starting with the Light-Matrix, the top left-hand matrix in (3.1.3), the first line from the top, $C_{\infty}:[Pr, Po, K, H]$, specifies the fundamental architecture of quantum particles. As explored previously in this chapter, Quarks are an emergence of Light's property of Knowledge, Leptons are an emergence of Light's property of Power, Bosons are an emergence of Light's property of Harmony, and the Higgs-boson is an emergence of Light's property of Presence. The fundamental architecture of these aspects hence, is an emergence of the properties of Light at ∞.

Line 3 in the Light-Matrix, $C_K: [S_{Pr}, S_{Po}, S_K, S_H]$, elaborates the sets for Presence, Power, Knowledge, and Harmony, each containing multiple elements. For example, as explored previously in the section on Quarks, various elements derived from the four sets would define the behavior of quarks and could be functions such as 'composite-arrangements', 'specifying attributes' amongst others, hence collectively describing quarks' way of being. Specifically, Line 5, $C_{N:} f(S_{Pr}\ x\ S_{Po}\ x\ S_K\ x\ S_H)$, suggests that unique seeds are created from a combination of such elements from all four sets, with a particular element leading, that in effect creates the distinctness possible at the level of quantum particles.

Line 6, $(\downarrow R_{C_U} = f(R_{C_N}))$, specifies quantization between the layer where the seeds are formed, and the physical layer we are familiar with, and as explored in Chapter 3.5 and 3.6, will result in Line 7, C_U: $[P,V,M,C]$, hence changing the material fabric of existence. The possibilities represented by Lines 1 through 5 hence concretize through the quantization represented by Line 6 to become the quantum particles with its physical (related to Presence), vital (related to Power), mental (related to Knowledge), and connection (related to Harmony) aspects now existing in material reality typified by Light moving at c. Note that just as Line 6 represents a process of quantization relating the layer of reality created by Light traveling at c with the antecedent layers, so too Lines 2 and 4 as previously discussed, also represent quantization of a more subtle kind that ultimately plays a critical part in allowing the material fabric to express infinite diversity.

Typically it is the process as captured by the Space-Matrix that will determine if Line 6 is activated. Specifically patterns at the untransformed layer, U, will need to be overcome, as specified by the second-line from the bottom of the Space-Matrix: $(\uparrow > P_x)$. But as specified by the bottom-line of the Time-Matrix, reproduced below, it is only with the advent of the human-system that the automaticity of the action of meta-levels is reversed:

$$U \rightarrow \begin{matrix} t \sim E_{Cell}; TC: M_3 \rightarrow U \\ t \sim E_{Human}; TC: U \rightarrow M_3 \end{matrix}$$

Hence in the case of the quantum particle system, which in this emergence is a pre-human system, the fact that patterns do not need to be overcome means that quantization happens automatically.

Nonetheless, and given this context it is useful to review
Equation 3.6.5, Quantization Effect of Organization on Material Fabric:

Impact on Materal Fabric

$$= \begin{bmatrix} |[L][S][T]TC \rightarrow x_T|_{\langle x_U|x_T\rangle} \\ \times \\ \left(|mod(Z_Q)|_{Y>U}\right) \\ \ni \\ Z_Q, Z \in \mathbb{U}\,(Space, Time, Energy, Gravity) \end{bmatrix}$$

Line 1 in the matrix is simply (3.6.1) the Simplified Light-Space-Time Emergence equation. To understand if quantization is going to occur, the organization resulting from the application of Line 1 is, where 'is' means 'multiplied' or ($\times$), by a 'modulation', specified by (mod), of a fourfold space-time-energy-gravity quantization (Z_Q) kind, so long as Y>U, that is, the operative bases are relatively transformed and hence have an active influence greater than U. The fourfold quantization is specified by ($\ni Z_Q, Z \in U\ (Space, Time, Energy, Gravity)$) where ($\ni$) the quantization being applied (Z) is each (U) of the elements of the set (Space, Time, Energy, Gravity).

But as just summarized in the Time-Matrix in (3.1.3) Y is by definition greater than U and hence quantization is automatic. In terms of quantum particles such quantization implies that wholeness becomes fully active through specific space, time, energy, and gravity quantization to create an holistic "ecosystem" with its own "quantum particle logic" as it were. The wholeness has now precipitated into the material fabric and is available to be consciously and unconsciously tapped into.

Note that while quantum particles as one of the earlier material manifestations of Light exists pervasively, technologies that leverage its deeper or implicit aspects, such as the intent of various kinds of quantum particles, will need the full operation of the Time-Matrix starting with the overcoming of patterns at U by human systems, in order to be fully leveraged. Such dynamics that in a sense 'evolve' the functioning of the quantum particle system will be reviewed in more detail in Section 10.

Chapter 5.2: Bosons

The bosons as mentioned can be thought of as force-carriers and allow all known matter particles to interact.

But when we look at bosons in more detail there are three fundamental bosons – the photon, the W and Z bosons, the gluon - and one hypothetical boson – the graviton.

Photon

The photon is the carrier particle of the electromagnetic force. But in the scheme of things the electromagnetic force pervades everything and as explored in the section on the electromagnetic spectrum appears to be foundational to this reality we are in. So it could be said that it is related to Presence or is an emergence of Light's property of Presence at the level of quantum force-carriers.

Hence, an equation for the photon, Equation 5.2.1 could be:

$$Sig_{photon} = Xa + \overline{Yb_{0-n}} \quad where \begin{bmatrix} X \in [S_{System_{Pr}}] \\ Y \in [S_{System_{Pr}}, S_{System_{P}}, S_{System_{K}}, S_{System_{N}}] \\ a, b \text{ are integers}; a > b \end{bmatrix}$$

Eq 5.2.1: Signature of Photon

While the primary element X, would continue to be the same as X for bosons, as in Equation 5.1.3, there will be additional secondary elements Y that will further qualify the attributes or functionality of photons. As an example such elements could be 'pervasiveness', 'multiple wave handler', amongst others.

Gluon

The gluon is the carrier particle of what is known as the strong nuclear force and holds quarks together in their inherently composite arrangements. But we have posited that quarks are related to or are an emergence of Knowledge at the quantum-particle level. Hence it must be that the gluon is an emergence of Light's property of Knowledge at the level of quantum force-carriers.

Hence, Equation 5.2.2 is an equation for the gluon:

$$Sig_{gluon} = Xa + \overline{Yb_{0-n}} \quad where \begin{bmatrix} X \in [S_{System_K}] \\ Y \in [S_{System_{Pr}}, S_{System_P}, S_{System_K}, S_{System_N}] \\ a, b \ are \ integers; a > b \end{bmatrix}$$

Eq 5.2.2: Signature of Gluon

The primary element X would continue to be the same as in the equation for the boson. There are likely additional secondary elements Y such as 'concentrated', 'intense connection', amongst others that would collectively determine the behavior of gluons.

W and Z Bosons

The W and Z bosons are the carrier particle for the weak interactions, responsible for the decay of massive quarks and leptons into lighter quarks and leptons. But this usually is accompanied by the release of energy and power and so it must be that W and Z bosons are an emergence of Light's property of Power at the level of quantum force-carriers.

Hence, Equation 5.2.3, for W and Z bosons is as follows:

$$Sig_{W,Z\ bosons} = Xa + \overline{Yb_{0-n}} \quad where \begin{bmatrix} X \in [S_{System_P}] \\ Y \in [S_{System_{Pr}}, S_{System_P}, S_{System_K}, S_{System_N}] \\ a, b \ are \ integers; a > b \end{bmatrix}$$

Eq 5.2.3: Signature of W and Z Bosons

The primary element X would continue to be the same as in the equation for the boson. There are likely additional secondary elements Y such as 'weak link', 'release of energy', amongst others that would collectively determine the behavior of W and Z bosons.

Graviton

This leaves the hypothetical graviton that is thought to be the carrier particle for the force of gravity. But gravity is what holds astronomical objects together in relationship. Hence it must be that gravitons are an emergence of Light's property of Harmony at the level of quantum force-carriers.

Hence, Equation 5.2.4, for the hypothetical graviton is as follows:

$$Sig_{graviton} =$$

$$Xa + \overline{Yb_{0-n}} \quad where \begin{bmatrix} X \in [S_{System_N}] \\ Y \in [S_{System_{P_T}}, S_{System_P}, S_{System_K}, S_{System_N}] \\ a, b \text{ are integers}; a > b \end{bmatrix}$$

Eq 5.2.4: Signature of Graviton

The primary element X would continue to be the same as in the equation for the boson. There are likely additional secondary elements Y such as 'mass curving space', 'space defining object movement', amongst others that would collectively determine the behavior of gravitons.

So we see that the underlying properties of light - Presence, Knowledge, Power, and Harmony - emerge as photons, gluons, W and Z bosons, and the hypothetical graviton at the quantum force-carrier level.

Light's Emergence as Bosons

As discussed previously the Light-Space-Time Emergence equation (3.1.3) being iterative, can be used to model emergence as it proceeds from simpler four-fold to more complex four-fold manifestations. Hence (3.1.3) has already been applied to suggest the emergence of the electromagnetic spectrum and quantum particles in general. Here it will be applied to suggest the emergence of a sub-class of quantum particles, Bosons. But further, as implied by (3.6.5), the Quantization Effect of Organization on Material Fabric equation, any process of organization, such as the architecture and cohesiveness of bosons when viewed with the lens provided here, has the possibility of altering the material fabric or fabric of existence so long as the bases involved are driven primarily by a meta-level.

As suggested by (3.1.3), reproduced below for convenience, the architecture and details of bosons can be seen to be the result of the application of the Light, Space, and Time matrices as will be elaborated:

$$
Emergence_{light-space-time} = \left[\begin{array}{cc}
\begin{bmatrix}
C_{\infty}: [Pr, Po, K, H] \\
(\downarrow R_{C_K} = f(R_{C_{\infty}})) \\
C_K: [S_{Pr}, S_{Po}, S_K, S_H] \\
(\downarrow R_{C_N} = f(R_{C_K})) \\
C_{N:} f(S_{Pr}\; x\; S_{Po}\; x\; S_K\; x\; S_H) \\
(\downarrow R_{C_U} = f(R_{C_N})) \\
C_U: [P, V, M, C]
\end{bmatrix}_{Light}
&
\begin{bmatrix}
M_3 \rightarrow System_x \\
(\uparrow F \rightarrow I) \\
M_2 \rightarrow S_{System\,x} \\
(\uparrow Sig \rightarrow F) \\
M_1 \rightarrow Sig_x \\
(\uparrow > P_x) \\
U \rightarrow x_U
\end{bmatrix}_{Space}
\\
\begin{bmatrix}
M_3 : -\infty \leq t \leq \infty \\
\downarrow \\
M_2 : 0 \geq t > \infty \\
\downarrow \\
M_1 : 0 > t > \infty \\
\downarrow \\
U \rightarrow \begin{array}{l} t \sim E_{Cell}; \text{TC}: M_3 \rightarrow U \\ t \sim E_{Human}; \text{TC}: U \rightarrow M_3 \end{array}
\end{bmatrix}_{Time}
&
TC \rightarrow x_T
\end{array}\right]_{\langle x_U | x_T \rangle}
$$

Starting with the Light-Matrix, the top left-hand matrix in (3.1.3), the first line from the top, $C_{\infty}:[Pr, Po, K, H]$, specifies the fundamental architecture of bosons. As explored previously in this chapter, Gluons are an emergence of Light's property of Knowledge, W and Z bosons are an emergence of Light's property of Power, Photons are an emergence of Light's property of Presence, and the Graviton is an emergence of Light's property of Harmony. The fundamental architecture of these aspects hence, is an emergence of the properties of Light at ∞.

Line 3 in the Light-Matrix, $C_K: [S_{Pr}, S_{Po}, S_K, S_H]$, elaborates the sets for Presence, Power, Knowledge, and Harmony, each containing multiple elements. For example, as explored previously in the section on photons, various elements derived from the four sets would define the behavior of photons and could be functions such as 'pervasiveness', 'multiple wave handler', amongst others, hence collectively describing photons' way of being. Specifically, Line 5, $C_{N:} f(S_{Pr}\; x\; S_{Po}\; x\; S_K\; x\; S_H)$, suggests that unique seeds are created from a combination of such elements from all four sets, with a particular element leading, that in effect creates the distinctness possible at the level of bosons.

Line 6, $(\downarrow R_{C_U} = f(R_{C_N}))$, specifies quantization between the layer where the seeds are formed, and the physical layer we are familiar with, and as explored in Chapter 3.5 and 3.6, will result in Line 7, C_U: $[P,V,M,C]$, hence changing the material fabric of existence. The possibilities represented by Lines 1 through 5 hence concretize through the quantization represented by Line 6 to become the bosons with its physical (related to Presence), vital (related to Power), mental (related to Knowledge), and connection (related to Harmony) aspects now existing in material reality typified by Light moving at c. Note that just as Line 6 represents a process of quantization relating the layer of reality created by Light traveling at c with the antecedent layers, so too Lines 2 and 4 as previously discussed, also represent quantization of a more subtle kind that ultimately plays a critical part in allowing the material fabric to express infinite diversity.

Typically it is the process as captured by the Space-Matrix that will determine if Line 6 is activated. Specifically patterns at the untransformed layer, U, will need to be overcome, as specified by the second-line from the bottom of the Space-Matrix: $(\uparrow > P_x)$. But as specified by the bottom-line of the Time-Matrix, reproduced below, it is only with the advent of the human-system that the automaticity of the action of meta-levels is reversed:

$$U \rightarrow \begin{matrix} t \sim E_{Cell};TC: M_3 \rightarrow U \\ t \sim E_{Human};TC: U \rightarrow M_3 \end{matrix}$$

Hence in the case of bosons, which in this emergence is a pre-human system, the fact that patterns do not need to be overcome means that quantization happens automatically.

Nonetheless, and given this context it is useful to review
Equation 3.6.5, Quantization Effect of Organization on Material Fabric:

Impact on Materal Fabric

$$= \begin{bmatrix} |[L][S][T]TC \rightarrow x_T|_{\langle x_U | x_T \rangle} \\ \times \\ \left(|mod(Z_Q)|_{Y>U}\right) \\ \ni \\ Z_Q, Z \in \mathbb{U}\ (Space, Time, Energy, Gravity) \end{bmatrix}$$

Line 1 in the matrix is simply (3.6.1) the Simplified Light-Space-Time Emergence equation. To understand if quantization is going to occur, the organization resulting from the application of Line 1 is, where 'is' means 'multiplied' or ($\times$), by a 'modulation', specified by (mod), of a fourfold space-time-energy-gravity quantization (Z_Q) kind, so long as Y>U, that is, the operative bases are relatively transformed and hence have an active influence greater than U. The fourfold quantization is specified by ($\ni Z_Q, Z \in U\ (Space, Time, Energy, Gravity)$) where ($\ni$) the quantization being applied (Z) is each (U) of the elements of the set (Space, Time, Energy, Gravity).

But as just summarized in the Time-Matrix in (3.1.3) Y is by definition greater than U and hence quantization is automatic. In terms of bosons such quantization implies that wholeness becomes fully active through specific space, time, energy, and gravity quantization to create an holistic "ecosystem" with its own "boson logic" as it were. The wholeness has now precipitated into the material fabric and is available to be consciously and unconsciously tapped into.

Note that while bosons as one of the earlier material manifestations of Light exists pervasively, technologies that leverage its deeper or implicit aspects, such as the intent of various bosons, will need the full operation of the Time-Matrix starting with the overcoming of patterns at U by human systems, in order to be fully leveraged. Such dynamics that in a sense 'evolve' the functioning of the boson system will be reviewed in more detail in Section 10.

Chapter 5.3: Atoms

Quantum particles in turn create atoms, and all known atoms can be classified into one of four groups: The s-group, p-group, d-group, and f-group. But what are these groups and are they too emergences of Light as it continues its journey of crystalizing the possibility or potentiality or fourfold functionality within it?

s-Group

The s-group consists primarily of what are known as alkali metals and alkali earth metals. These alkali metals are known to easily lose electrons and form what is known as positive ions. When they lose electrons energy is gained, but when the electrons are taken up by other atoms in proximity there is a lot of energy released. Some have referred to these groups as "violent worlds" (Tweed, 2003), and it has been pointed out that stars shine because they are transmuting vast amount of hydrogen into helium, both of which are s-group elements. So one gets the sense that the s-group may be an emergence of Light's property of Power.

Stars and suns are also known to be the crucibles where all the different kinds of atoms are created. So these furnaces of power by virtue of their heat and high pressure are able to force electrons and protons and neutrons to come together to create all the different types of atoms known in the universe.

But philosophically what are the s-group atoms or elements? These are atoms where there is an equal likelihood of an electron being anywhere in a symmetrical sphere around the nucleus. The other groups are all similarly defined by likelihoods of electrons being within a possible pattern around the nucleus. The patterns that distinguishes s-group atoms is a sphere, and since all other patterns can be thought of as occurring within some sphere, in some sense this is like an imprint or precipitation or emergence that allows other kinds of emergences to surface within it. So the elements that are part of the s-group may be thought of as the adventurers with courage who venture into a brave new world to create some foundation by which all other element-creations can follow.

The fact that hydrogen and helium are known to constitute 98% of the Universe (Heiserman, 1991) relative to other elements therefore makes sense in this view, especially since hydrogen and helium provide the fuel with which the star-furnaces manufacture all other elements.

So s-group elements seem to embody functions such as power, energy, adventure, courage, and can be thought of as an emergence of the property of Light to do with Power.

Hence, a series of equations linked to S_{System_P} as the prime set can be suggested, starting with the s-group mapping, as in Equation 5.3.1:

$$Element_{s-group} = Xa + \overline{Yb_{0-n}} \quad where \begin{bmatrix} X \in [S_{System_P}] \\ Y \in [S_{System_{Pr}}, S_{System_P}, S_{System_K}, S_{System_N}] \\ a, b \text{ are integers}; a > b \end{bmatrix}$$

Eq 5.3.1: s-Group Element

Further, the equivalent mapping between traditional element groupings and S_{System_P} as in Equations 5.3.2 and 5.3.3 can be specified:

$$Element_{Alkali\ metal} = Xa + \overline{Yb_{0-n}} \quad where \begin{bmatrix} X \in [S_{System_P}] \\ Y \in [S_{System_{Pr}}, S_{System_P}, S_{System_K}, S_{System_N}] \\ a, b \text{ are integers}; a > b \end{bmatrix}$$

Eq 5.3.2: Alkali Metal Element

$$Element_{Alkali\ earth\ metal} = Xa + \overline{Yb_{0-n}} \quad where \begin{bmatrix} X \in [S_{System_P}] \\ Y \in [S_{System_{Pr}}, S_{System_P}, S_{System_K}, S_{System_N}] \\ a, b \text{ are integers}; a > b \end{bmatrix}$$

Eq 5.3.3: Alkali Earth Metal Element

Further, as a representative element belonging to the Alkali Metal group the equation for Lithium (Li), as in Equation 5.3.4, would be:

$$Element_{Lithium} = Xa + \overline{Yb_{0-n}} \quad where \begin{bmatrix} X \in [S_{System_P}] \\ Y \in [S_{System_{Pr}}, S_{System_P}, S_{System_K}, S_{System_N}] \\ a, b \text{ are integers}; a > b \end{bmatrix}$$

Eq 5.3.4: Lithium

Note that the primary element X in all these cases may be a function or attribute along the lines of 'expresses power'. The secondary elements Y in all these would have some elements being the same, and then would have specific differences as one gets into sub-classes and the individual elements themselves.

p-Group

Atoms or elements belonging to the p-group are those with the likelihood of electrons occurring equally on either side of the nucleus, like a dumbbell of sorts.

There are some very significant elements in this group that are part of the metal, metalloid, non-metal, halogen, and noble gas sub-groupings. Carbon, Nitrogen, Oxygen, and Silicon are some of the sample elements. Looking at the types of elements present in this group it is as though all the element possibilities have been represented within it. It is perhaps that the possibility of ideas behind all elements has emerged in this group and one can hypothesize that this group may be a reflection of the property of Knowledge, forming archetypes from which all other elements are created.

Philosophically, the one spherical shell or probability cloud (s) becoming two shells like a dumbbell (p) signifies the creation of an essential polarity within a unit space. So in a sense the one becoming two is the first instance of variability in space. The two, spaced in 3-dimensions around the nucleus in a sense create six switches becoming an attractor or allowing a vaster number of different kinds of elements to surface. So there is a sense of the 'idea' of the element that comes into focus.

But further, the essential elements that allow both thinking and virtual thinking machines to come into being, are also contained within this group. Carbon is the basis of DNA and of all life. The fact the Silicon, directly below it in the periodic table and therefore sharing essential qualities, is considered the basis of all virtual thinking machines is therefore perhaps significant and may reinforce the notion that the p-group is a precipitation of Light's property of Knowledge.

Hence, a series of equations linked to S_{System_K} as the prime set can be suggested, starting with the p-group mapping, as in Equation 5.3.5:

$$Element_{\,p-group} =$$

$$Xa + \overline{Yb_{0-n}} \quad where \left[\begin{array}{c} X \in [S_{System_K}] \\ Y \in [S_{System_{Pr}}, S_{System_p}, S_{System_K}, S_{System_N}] \\ a, b\ are\ integers; a > b \end{array} \right]$$

Eq 5.3.5: p-group Element

Further, the equivalent mapping between traditional element groupings and S_{System_K} as in Equations 5.3.6 through 5.3.11 can also be specified:

$$Element_{\,Metal} =$$

$$Xa + \overline{Yb_{0-n}} \quad where \left[\begin{array}{c} X \in [S_{System_K}] \\ Y \in [S_{System_{Pr}}, S_{System_p}, S_{System_K}, S_{System_N}] \\ a, b\ are\ integers; a > b \end{array} \right]$$

Eq 5.3.6: Metal Element

$$Element_{\,Metalloid} =$$

$$Xa + \overline{Yb_{0-n}} \quad where \left[\begin{array}{c} X \in [S_{System_K}] \\ Y \in [S_{System_{Pr}}, S_{Syste\square_p}, S_{System_K}, S_{System_N}] \\ a, b\ are\ integers; a > b \end{array} \right]$$

Eq 5.3.7: Metalloid Element

$$Element_{Non-Metal} = Xa + \overline{Yb_{0-n}} \quad where \begin{bmatrix} X \in [S_{System_K}] \\ Y \in [S_{System_{P_T}}, S_{System_P}, S_{System_K}, S_{System_N}] \\ a, b \text{ are integers}; a > b \end{bmatrix}$$

Eq 5.3.8: Non-Metal Element

$$Element_{Halogen} = Xa + \overline{Yb_{0-n}} \quad where \begin{bmatrix} X \in [S_{System_K}] \\ Y \in [S_{System_{P_T}}, S_{System_P}, S_{System_K}, S_{System_N}] \\ a, b \text{ are integers}; a > b \end{bmatrix}$$

Eq 5.3.9: Halogen Element

$$Element_{Noble\ Gas} = Xa + \overline{Yb_{0-n}} \quad where \begin{bmatrix} X \in [S_{System_K}] \\ Y \in [S_{System_{P_T}}, S_{System_P}, S_{System_K}, S_{System_N}] \\ a, b \text{ are integers}; a > b \end{bmatrix}$$

Eq 5.3.10: Noble Gas Element

Further, as a representative element belonging to the Non-Metal group the equation for Carbon (C), as in Equation 5.3.11, would be:

$$Element_{Carbon} = Xa + \overline{Yb_{0-n}} \quad where \begin{bmatrix} X \in [S_{System_K}] \\ Y \in [S_{System_{P_T}}, S_{System_P}, S_{System_K}, S_{System_N}] \\ a, b \text{ are integers}; a > b \end{bmatrix}$$

Eq 5.3.11: Carbon

Note that the primary element X in all these cases may be a function or attribute along the lines of 'knowledge'. The secondary elements Y in all these would have some elements being the same, and then would have specific differences as one gets into sub-classes and the individual elements themselves.

d-Group

The d-group comprises the Transition Metals. These metals are generally hard and strong, exhibit corrosive resistance, and can be thought of as workhorse elements. Many industrial and well-known elements sit in this group: Titanium, Chromium, Manganese, Iron, Cobalt, Nickel, Copper, Zinc, Silver, Platinum, and Gold, amongst others.

The elements and atoms in the d-group are equally likely to show up in four possible lobes or probability-spaces around the nucleus. Four lobes occurring in multiple possible planes around the nucleus will likely create a space of stability, since there is a possibility of four lobes creating the four vertices of a tetrahedron (Fuller, 1982) that has been positioned as one of the most stable shapes in the universe.

Much of the constructed world around us is created from these elements. Further, most of the series in the group easily lose one or more electrons thereby easily combining with other atoms to form a vast array of compounds. Also, looking more broadly at the function of these elements, it can be seen that these metals exist for service, to help bring about perfection in the constructed world, to help much of the machinery in which they are used, and to assist the processes dependent on them to be completed with diligence. Hence, these transition metals appear to be an emergence of Light's property of Presence.

Therefore, a series of equations linked to $S_{System_{Pr}}$ as the prime set can be suggested, starting with the d-group mapping, as in Equation 5.3.12:

$$Element_{\,d-group} =$$

$$Xa + \overline{Yb_{0-n}} \ \ where \begin{bmatrix} X \in [S_{System_{Pr}}] \\ Y \in [S_{System_{Pr}}, S_{System_{P}}, S_{System_{K}}, S_{System_{N}}] \\ a, b \ are \ integers; a > b \end{bmatrix}$$

Eq 5.3.12: d-group Element

Further, the equivalent mapping between traditional element groupings and $S_{System_{Pr}}$, as in Equation 5.3.13, can also be specified:

$$Element_{Transition\ Metal} = Xa + \overline{Yb_{0-n}} \ \ where \begin{bmatrix} X \in [S_{System_{Pr}}] \\ Y \in [S_{System_{Pr}}, S_{System_{P}}, S_{System_{K}}, S_{System_{N}}] \\ a, b\ are\ integers; a > b \end{bmatrix}$$

Eq 5.3.13: Transition Metal Element

Further, as a representative element belonging to the Transition-Metal group the equation for Gold (Au), as in Equation 5.3.14, would be:

$$Element_{Gold} = Xa + \overline{Yb_{0-n}} \ \ where \begin{bmatrix} X \in [S_{System_{Pr}}] \\ Y \in [S_{System_{Pr}}, S_{System_{P}}, S_{System_{K}}, S_{System_{N}}] \\ a, b\ are\ integers; a > b \end{bmatrix}$$

Eq 5.3.14: Gold

Note that the primary element X in all these cases may be a function or attribute along the lines of 'work horse'. The secondary elements Y in all these would have some elements being the same, and then would have specific differences as one gets into sub-classes and the individual elements themselves.

f-Group

The f-group comprises of the Lanthanides and Actinides. Philosophically, elements in the f-group consist of 6 lobes around the nucleus within which an electron may be found. 6 lobes will exist in multiple planes around the nucleus and suggests the notion of extended relationship and collectivity: the attempt to build larger and larger bonds within a small space. Considering this it is likely that the f-group is an emergence of Light's property of Harmony.

Thinking about Lanthanides, some interesting facts may reinforce this notion. First, the spin of electrons in a lanthanides' outer shell is aligned, creating a strong magnetic field. The notion of creating a strong magnetic field seems to be consistent with the notion of engendering a collectivity through the ordered attraction and repulsion of elements. Second, these elements curiously occur together in nature often in the same ores (Gray, 2009) and are chemically interchangeable also suggesting the notion of forming a tight intra-group collectivity.

Actinides on the other hand are inherently radioactive. This implies that these elements have inherently crossed a threshold of stability and have the urge, over their own half-lives, to decompose into other elements. This natural urge may suggest some boundary conditions on the notion of collectivity and nurturing, giving additional insight into the nature of collectivity and nurturing. Further, the entire actinide group, as opposed to the lanthanide group that is inherently stable, is unstable. It is curious that both these should be part of the f-group, and they must provide insight into boundary conditions into the notion of collectivity in elements.

Hence, a series of equations linked to S_{System_N} as the prime set can be suggested, starting with the f-orbital mapping, as in Equation 5.3.15:

$$Element_{f-group} = Xa + \overline{Yb_{0-n}} \quad where \begin{bmatrix} X \in [S_{System_N}] \\ Y \in [S_{System_{Pr}}, S_{System_P}, S_{System_K}, S_{System_N}] \\ a, b \text{ are integers}; a > b \end{bmatrix}$$

Eq 5.3.15: f-group Element

Further, the equivalent mapping between traditional element groupings and S_{System_N}, as in Equations 5.3.16 and 5.3.17, can also be specified:

$$Element_{Lanthanide} = Xa + \overline{Yb_{0-n}} \quad where \begin{bmatrix} X \in [S_{System_N}] \\ Y \in [S_{System_{Pr}}, S_{System_P}, S_{System_K}, S_{System_N}] \\ a, b \text{ are integers}; a > b \end{bmatrix}$$

Eq 5.3.16: Lanthanide Element

$$Element_{Actinide} = Xa + \overline{Yb_{0-n}} \quad where \left[\begin{array}{c} X \in [S_{System_N}] \\ Y \in [S_{System_{P_Y}}, S_{System_P}, S_{System_K}, S_{System_N}] \\ a, b \; are \; integers; a > b \end{array} \right]$$

Eq 5.3.17: Actinide Element

Further, as a representative element belonging to the Lanthanide group the equation for Lanthanum (La), as in Equation 5.3.18, would be:

$$Element_{Lanthanum} = Xa + \overline{Yb_{0-n}} \quad where \left[\begin{array}{c} X \in [S_{System_N}] \\ Y \in [S_{System_{P_Y}}, S_{System_P}, S_{System_K}, S_{System_N}] \\ a, b \; are \; integers; a > b \end{array} \right]$$

Eq 5.3.18: Lanthanum

Note that the primary element X in all these cases may be a function or attribute along the lines of 'experiments in collectivity'. The secondary elements Y in all these would have some elements being the same, and then would have specific differences as one gets into sub-classes and the individual elements themselves.

Light's Emergence as Atoms

As discussed previously the Light-Space-Time Emergence equation (3.1.3) being iterative, can be used to model emergence as it proceeds from simpler four-fold to more complex four-fold manifestations. Hence (3.1.3) has already been applied to suggest the emergence of the electromagnetic spectrum, quantum particles in general, and bosons as a further instance of a particular kind of quantum particle. Here it will be applied to suggest the emergence of Atoms. But further, as implied by (3.6.5), the Quantization Effect of Organization on Material Fabric equation, any process of organization, such as the architecture and cohesiveness of atoms, has the possibility of altering the material fabric or fabric of existence so long as the bases involved are driven primarily by a meta-level.

As suggested by (3.1.3), reproduced below for convenience, the architecture and details of atoms can be seen to be the result of the application of the Light, Space, and Time matrices as will be elaborated:

$$
Emergence_{light-space-time} = \left| \begin{matrix} \begin{bmatrix} C_{\infty}: [Pr, Po, K, H] \\ (\downarrow R_{C_K} = f(R_{C_{\infty}})) \\ C_K: [S_{Pr}, S_{Po}, S_K, S_H] \\ (\downarrow R_{C_N} = f(R_{C_K})) \\ C_{N}: f(S_{Pr}\ x\ S_{Po}\ x\ S_K\ x\ S_H) \\ (\downarrow R_{C_U} = f(R_{C_N})) \\ C_U: [P, V, M, C] \end{bmatrix}_{Light} & \begin{bmatrix} M_3 \rightarrow System_X \\ (\uparrow F \rightarrow I) \\ M_2 \rightarrow S_{System X} \\ (\uparrow Sig \rightarrow F) \\ M_1 \rightarrow Sig_x \\ (\uparrow > P_x) \\ U \rightarrow x_U \end{bmatrix}_{Space} \\ \begin{bmatrix} & M_3 : -\infty \leq t \leq \infty \\ & \downarrow \\ & M_2 : 0 \geq t > \infty \\ & \downarrow \\ & M_1 : 0 > t > \infty \\ & \downarrow \\ U \rightarrow & \begin{matrix} t \sim E_{Cell}; TC: M_3 \rightarrow U \\ t \sim E_{Human}; TC: U \rightarrow M_3 \end{matrix} \end{bmatrix}_{Time} & TC \rightarrow x_T \end{matrix} \right|_{\langle x_U | x_T \rangle}
$$

Starting with the Light-Matrix, the top left-hand matrix in (3.1.3), the first line from the top, $C_{\infty}:[Pr, Po, K, H]$, specifies the fundamental architecture of atoms. As explored previously in this chapter, p-Group atoms an emergence of Light's property of Knowledge, s-Group atoms are an emergence of Light's property of Power, d-Group atoms are an emergence of Light's property of Presence, and the f-Group atoms are an emergence of Light's property of Harmony. The fundamental architecture of these aspects hence, is an emergence of the properties of Light at ∞.

Line 3 in the Light-Matrix, $C_K: [S_{Pr}, S_{Po}, S_K, S_H]$, elaborates the sets for Presence, Power, Knowledge, and Harmony, each containing multiple elements. For example, as explored previously in the section on s-Group, various elements derived from the four sets would define the behavior of s-Group atoms and could be functions such as 'power', 'energy', 'adventure', and 'courage', amongst others, hence collectively describing s-Groups atoms' way of being. Specifically, Line 5, $C_{N}: f(S_{Pr}\ x\ S_{Po}\ x\ S_K\ x\ S_H)$, suggests that unique seeds are created from a combination of such elements from all four sets, with a particular element leading, that in effect creates the distinctness possible at the level of atoms.

Line 6, $(\downarrow R_{C_U} = f(R_{C_N}))$, specifies quantization between the layer where the seeds are formed, and the physical layer we are familiar with, and as explored in Chapter 3.5 and 3.6, will result in Line 7, C_U: $[P,V,M,C]$, hence changing the material fabric of existence. The possibilities represented by Lines 1 through 5 hence concretize through the quantization represented by Line 6 to become the atoms with its physical (related to Presence), vital (related to Power), mental (related to Knowledge), and connection (related to Harmony) aspects now existing in material reality typified by Light moving at c. Note that just as Line 6 represents a process of quantization relating the layer of reality created by Light traveling at c with the antecedent layers, so too Lines 2 and 4 as previously discussed, also represent quantization of a more subtle kind that ultimately plays a critical part in allowing the material fabric to express infinite diversity.

Typically it is the process as captured by the Space-Matrix that will determine if Line 6 is activated. Specifically patterns at the untransformed layer, U, will need to be overcome, as specified by the second-line from the bottom of the Space-Matrix: $(\uparrow > P_x)$. But as specified by the bottom-line of the Time-Matrix, reproduced below, it is only with the advent of the human-system that the automaticity of the action of meta-levels is reversed:

$$U \rightarrow \begin{array}{l} t \sim E_{Cell};TC: M_3 \rightarrow U \\ t \sim E_{Human};TC: U \rightarrow M_3 \end{array}$$

Hence in the case of atoms, which in this emergence is a pre-human system, the fact that patterns do not need to be overcome means that quantization happens automatically.

Nonetheless, and given this context it is useful to review
Equation 3.6.5, Quantization Effect of Organization on Material Fabric:

Impact on Materal Fabric

$$= \begin{bmatrix} \|[L][S][T]TC \rightarrow x_T|_{\langle x_U|x_T\rangle} \\ \times \\ \left(|mod(Z_Q)|_{Y>U}\right) \\ \ni \\ Z_Q, Z \in \mathbb{U}\ (Space, Time, Energy, Gravity) \end{bmatrix}$$

Line 1 in the matrix is simply (3.6.1) the Simplified Light-Space-Time Emergence equation. To understand if quantization is going to occur, the organization resulting from the application of Line 1 is, where 'is' means 'multiplied' or ($\times$), by a 'modulation', specified by (mod), of a fourfold space-time-energy-gravity quantization (Z_Q) kind, so long as Y>U, that is, the operative bases are relatively transformed and hence have an active influence greater than U. The fourfold quantization is specified by ($\ni Z_Q, Z \in U\ (Space, Time, Energy, Gravity)$) where ($\ni$) the quantization being applied (Z) is each (U) of the elements of the set (Space, Time, Energy, Gravity).

But as just summarized in the Time-Matrix in (3.1.3) Y is by definition greater than U and hence quantization is automatic. In terms of atoms such quantization implies that wholeness becomes fully active through specific space, time, energy, and gravity quantization to create an holistic "ecosystem" with its own "atom logic" as it were. The wholeness has now precipitated into the material fabric and is available to be consciously and unconsciously tapped into.

Note that while atoms as one of the earlier material manifestations of Light exists pervasively, technologies that leverage its deeper or implicit aspects, such as the intent of various atoms, will need the full operation of the Time-Matrix starting with the overcoming of patterns at U by human systems, in order to be fully leveraged. Such dynamics that in a sense 'evolve' the functioning of the atom system will be reviewed in more detail in Section 10.

SECTION 6: EMERGENCE OF LIFE

This Section explores the emergence of Life through the Cell, complex human attributes such as thoughts and feelings, and uniqueness of individuality.

Hence Chapter 6.1 will explore the emergence and complexification of four-foldness through the primary molecular plans of nucleic acids, proteins, lipids, and polysaccharides and position the various combinations of this four-foldness as the basis of Natural History.

Chapter 6.2 will relate key human attributes of sensations, urges, desires, wills, feelings, emotions, and thought to the continuing journey of fourfold complexification.

Chapter 6.3 will relate truer individuality to the fourfold properties implicit in Light.

Chapter 6.1: Living Cells

The living cell has a universe of adaptability embedded in it. In 'The Machinery of Life', Goodsell, an Associate Professor of Molecular Biology at the Scripps Research Institute (Goodsell, 2010) suggests that every living thing on Earth uses a similar set of molecules to eat, to breathe, to move, and to reproduce. There are molecular machines that do the myriad things that distinguish living organisms that are identical in all living cells. This nanoscale machinery of cells uses four basic molecular plans with unique chemical personalities: nucleic acids, proteins, lipids, and polysaccharides.

But further, it is a combination of these molecular plans that can cause specialization of a cell to create any feature of a plant or animal. At the level of plants for example, striking colors and scents can become the natural features of a plant. It can adapt its shape to allow various insects or bees to enter into it in different ways. It can store a myriad of nutrients in a myriad different ways. It can develop poisons to keep predatory animals away. At the level of animals for example, skins can toughen into hard shells. Feet can develop suction caps to allow mobility in a vertical direction. Eyes can develop broad vision, telescopic vision, or microscopic vision. The mechanics of all adaptability can be traced to the ability of the four molecular plans to coordinate and generate more granular cell-level features that give it some evolutionary advantage in the scheme of things. All Natural History, hence, can be traced to this ability of cells to specialize along a diverse array of specialized lines.

In this chapter we will examine the mathematics of the molecular plans that form the bases for such specialization to occur.

Nucleic Acids

Nucleic acids basically encode information. They store and transmit the genome, the hereditary information needed to keep the cell alive. They function as the cell's librarians and contain information on how to make proteins and when to make them.

They are hence, the keepers of a cell's knowledge, its wisdom, its ability to make laws, the vehicle to spread knowledge within cells and to the next generation of cells. Being so, one can see that there is similarity with the set for system-knowledge highlighted earlier in Chapter 2.3. Reproducing Equation 2.3.3:

$$S_{System_K} \ni [Wisdom, Law\ Making, Spread\ of\ Knowledge ...]$$

Nucleic acids can therefore be thought of as a precipitation of system-knowledge at the cellular level.

Hence, a nucleic acid will have a generalized signature, as in Equation 6.1.1, derived from the system-knowledge family:

$$Sig_{nucleic\ acid} = Xa + \overline{Yb_{0-n}} \quad where \left[\begin{array}{c} X \in [S_{System_K}] \\ Y \in [S_{System_{Pr}}, S_{System_P}, S_{System_K}, S_{System_N}] \\ a, b\ are\ integers; a > b \end{array} \right]$$

Eq 6.1.1: Nucleic Acid

The primary element X could be an attribute or function such as 'keeper of knowledge'. Secondary elements Y could be 'protein laws', 'generational knowledge', amongst others. The collectivity of elements as per the equation would specify the character of a nucleic acid.

DNA and RNA, two types of nucleic acids, would hence have the equations as specified by Equation 6.1.2 and 6.1.3 respectively.

$$Sig_{DNA} = Xa + \overline{Yb_{0-n}} \quad where \left[\begin{array}{c} X \in [S_{System_K}] \\ Y \in [S_{System_{Pr}}, S_{System_P}, S_{System_K}, S_{System_N}] \\ a, b\ are\ integers; a > b \end{array} \right]$$

Eq 6.1.2: DNA

$$Sig_{RNA} = Xa + \overline{Yb_{0-n}} \quad where \left[\begin{array}{c} X \in [S_{System_K}] \\ Y \in [S_{System_{Pr}}, S_{System_P}, S_{System_K}, S_{System_N}] \\ a, b\ are\ integers; a > b \end{array} \right]$$

Eq 6.1.3: RNA

The primary element X in (6.1.2) and (6.1.3) would be the same as that for nucleic acids (6.1.1). The secondary elements Y however will be a larger and more specific set with many elements in common with (6.1.1).

Proteins

Proteins are the cells work-horses. Look anywhere in a cell and one will see proteins at work. Proteins are built in thousands of shapes and sizes, each performing a different function. As Goodsell describes, "some are built simply to adopt a defined shape, assembling into rods, nets, hollow spheres, and tubes. Some are molecular motors, using energy to rotate, or flex, or crawl. Many are chemical catalysts that perform chemical reactions atom-by-atom, transferring and transforming chemical groups exactly as needed." With their wide potential for diversity, proteins are constructed to perform most of the everyday tasks of the cells. In fact human cells build around 30,000 different kinds of proteins to execute on the diverse array of cellular level tasks.

Proteins hence, exist for service, to bring about perfection at the level of the cell, are characterized by extreme diligence and perseverance, and so on. Being so, one can see that there is similarity with the set for system-presence highlighted earlier in Chapter 2.3. Reproducing Equation 2.3.1:

$$S_{System_{Pr}} \ni [Service, Perfection, Diligence, Perseverance, \ldots]$$

Proteins can therefore be thought of as a precipitation of system-presence at the cellular level.

Hence, a protein could have a generalized signature, as in Equation 6.1.4, derived from the system-presence family:

$$Sig_{protein} = Xa + \overline{Yb_{0-n}} \quad where \begin{bmatrix} X \in [S_{System_{Pr}}] \\ Y \in [S_{System_{Pr}}, S_{System_{P}}, S_{System_{K}}, S_{System_{N}}] \\ a, b \text{ are integers}; a > b \end{bmatrix}$$

Eq 6.1.4: Protein

This could yield a vast number of functional proteins. In fact it may be possible that the 30,000 or so known proteins created by the human cell could each be specified by a signature equation of this nature. It may be possible to map existing proteins to functionality as suggested by the four sets of molecular plans.

Consider Insulin, for example. Insulin regulates the metabolism of carbohydrates, fats and protein by promoting the absorption of, especially, glucose from the blood into fat, liver and skeletal muscle cells. Equation 6.1.5 for Insulin would hence be:

$$Sig_{insulin} = Xa + \overline{Yb_{0-n}} \quad where \begin{bmatrix} X \in [S_{System_{Pr}}] \\ Y \in [S_{System_{Pr}}, S_{System_{P}}, S_{System_{K}}, S_{System_{N}}] \\ a, b \text{ are integers}; a > b \end{bmatrix}$$

Eq 6.1.5: Insulin

The primary element X could be an attribute or function such as 'workhorse. Secondary elements Y could be 'metabolic regulation, 'glucose absorption', 'blood to fat channel', amongst others. The collectivity of elements as per the equation would specify the character of insulin.

Consider Histones as another example. They are the chief protein components of chromatin, acting as spools around which DNA winds, and playing a role in gene regulation. Without histones, the unwound DNA in chromosomes would be very long (a length to width ratio of more than 10 million to 1 in human DNA). Equation 6.1.6 for Histones would hence be:

$$Sig_{histones} = Xa + \overline{Yb_{0-n}} \quad where \begin{bmatrix} X \in [S_{System_{Pr}}] \\ Y \in [S_{System_{Pr}}, S_{System_{P}}, S_{System_{K}}, S_{System_{N}}] \\ a, b \text{ are integers}; a > b \end{bmatrix}$$

Eq 6.1.6: Histones

The secondary element Y would have elements such as 'gene regulation', 'unwound DNA management', amongst others.

Lipids

Lipids by themselves are tiny molecules, but when grouped together form the largest structures of the cell. When placed in water, lipid molecules aggregate to form huge waterproof sheets. These sheets easily form boundaries at multiple levels and allow concentrated interactions and work to be performed within a cell. Hence, the nucleus and the mitochondria are contained within lipid-defined compartments. Similarly, each cell itself is contained within a lipid-defined boundary.

Lipids are therefore promoters of relationship, of harmony in the cell, of nurturing the cell-level division of labor, of allowing specialization and uniqueness to emerge, hence perhaps of earlier forms of compassion and love, and so on. The notion of such early forms of compassion is consistent with the biologist's perspective that at some point a gene for compassion was developed in pre-human species (Wright, 2009). Being so, one can see that there is similarity with the set for system-nurturing highlighted earlier in Chapter 2.3. Reproducing Equation 2.3.4:

$$S_{System_N} \ni [Love, Compassion, Harmony, Relationship \ldots]$$

This function of harmonization suggests that lipids can therefore be thought of as a precipitation of system-nurturing at the cellular level.

Lipids could have a generalized signature, as in Equation 6.1.7, derived from the system-nurturing family:

$$Sig_{lipid} = Xa + \overline{Yb_{0-n}} \quad where \begin{bmatrix} X \in [S_{System_N}] \\ Y \in [S_{System_{Pr}}, S_{System_P}, S_{System_K}, S_{System_N}] \\ a, b\ are\ integers; a > b \end{bmatrix}$$

Eq 6.1.7: Lipid

Specific lipids such as monoglycerides and phospholipids cold have the following equations:

$$Sig_{monoglyceride} = Xa + \overline{Yb_{0-n}} \quad where \begin{bmatrix} X \in [S_{System_N}] \\ Y \in [S_{System_{Pr}}, S_{System_P}, S_{System_K}, S_{System_N}] \\ a, b\ are\ integers; a > b \end{bmatrix}$$

Eq 6.1.8: Monoglyceride

$$Sig_{phospholipids} = Xa + \overline{Yb_{0-n}} \quad where \begin{bmatrix} X \in [S_{System_N}] \\ Y \in [S_{System_{Pr}}, S_{System_P}, S_{System_K}, S_{System_N}] \\ a, b\ are\ integers; a > b \end{bmatrix}$$

Eq 6.1.9: Phospholipids

The primary element X shared by each of the lipids could be an attribute or function such as 'compartmentalization'. Secondary elements Y could be of the nature of 'work breakdown', 'intra-cell love', amongst others, and would vary with each different kind of lipid.

Polysaccharides

Polysaccharides are long, often branched chains of sugar molecules. Sugars are covered with hydroxyl groups, which associate to form storage containers. As a result polysaccharides function as the storehouse of cell's energy. In addition polysaccharides are also used to build some of the most durable biological structures. The stiff shell of insects, for example are made of long polysaccharides.

Polysaccharides function to create energy, power, courage, strength thereby readying the cell for adventure, and so on. Being so, one can see that there is similarity with the set for system-power highlighted previously in Chapter 2.3. Reproducing Equation 2.3.2:

$$S_{System_p} \ni [Power, Courage, Adventure, Justice, ...]$$

Providing energy and strength, polysaccharides can be thought of as a precipitation of system-power at the cellular level.

Polysaccharides could have a generalized signature, as in Equation 6.1.10, derived from the system-power family:

$$Sig_{polysaccharide} =$$

$$Xa + \overline{Yb_{0-n}} \;\; where \begin{bmatrix} X \in [S_{Systemp}] \\ Y \in [S_{Systemp_{Pr}}, S_{Systemp}, S_{Systemp_K}, S_{Systemp_N}] \\ a, b \; are \; integers; a > b \end{bmatrix}$$

Eq 6.1.10: Polysaccharide

Glycogen is an example of a polysaccharide. Glycogen forms an energy reserve that can be quickly mobilized to meet a sudden need for glucose, but one that is less compact and more immediately available as an energy reserve than say triglycerides. Equation 6.1.11 for Glycogen follows:

$$Sig_{glycogen} =$$

$$Xa + \overline{Yb_{0-n}} \;\; where \begin{bmatrix} X \in [S_{Systemp}] \\ Y \in [S_{Systemp_{Pr}}, S_{Systemp}, S_{Systemp_K}, S_{Systemp_N}] \\ a, b \; are \; integers; a > b \end{bmatrix}$$

Eq 6.1.11: Glycogen

Cellulose is another example of a polysaccharide. Cellulose is a polymer made with repeated glucose units bonded together by beta-linkages. Humans and many animals lack an enzyme to break the beta-linkages, so they do not digest cellulose. Equation 6.1.12 for Cellulose follows:

$$Sig_{cellulose} = Xa + \overline{Yb_{0-n}} \quad where \begin{bmatrix} X \in [S_{System_P}] \\ Y \in [S_{System_{P_T}}, S_{System_P}, S_{System_K}, S_{System_N}] \\ a, b \text{ are integers}; a > b \end{bmatrix}$$

Eq 6.1.12: Cellulose

The primary element for the preceding polysaccharides X would be along the lines of 'energy storage'. The secondary elements may vary, with a Y element for Glycogen being 'rapid energy deployment' for example, and a Y element for Cellulose being 'bonded energy', for example.

Light's Emergence as Living Cells

As discussed previously the Light-Space-Time Emergence equation (3.1.3) being iterative, can be used to model emergence as it proceeds from simpler four-fold to more complex four-fold manifestations. Hence (3.1.3) has already been applied to suggest the emergence of the electromagnetic spectrum, quantum particles in general, bosons as a further instance of a particular kind of quantum particle, and atoms. Here it will be applied to suggest the emergence of Living Cells. But further, as implied by (3.6.5), the Quantization Effect of Organization on Material Fabric equation, any process of deeper organization, such as is responsible for the architecture and cohesiveness of living cells, has the possibility of altering the material fabric or fabric of existence so long as the bases involved are driven primarily by a meta-level.

As suggested by (3.1.3), reproduced below for convenience, the architecture and details of living cells can be seen to be the result of the application of the Light, Space, and Time matrices as will be elaborated:

$$
\begin{aligned}
&Emergence_{light-space-time} \\
&= \left\| \begin{array}{ll}
\begin{bmatrix} C_\infty: [Pr, Po, K, H] \\ (\downarrow R_{C_K} = f(R_{C_\infty})) \\ C_K: [S_{Pr}, S_{Po}, S_K, S_H] \\ (\downarrow R_{C_N} = f(R_{C_K})) \\ C_{N:}\, f(S_{Pr}\ x\ S_{Po}\ x\ S_K\ x\ S_H) \\ (\downarrow R_{C_U} = f(R_{C_N})) \\ C_U: [P, V, M, C] \end{bmatrix}_{Light} &
\begin{bmatrix} M_3 \rightarrow System_X \\ (\uparrow F \rightarrow I) \\ M_2 \rightarrow S_{System_X} \\ (\uparrow Sig \rightarrow F) \\ M_1 \rightarrow Sig_x \\ (\uparrow > P_x) \\ U \rightarrow x_U \end{bmatrix}_{Space} \\
\begin{bmatrix} M_3 : -\infty \le t \le \infty \\ \downarrow \\ M_2 : 0 \ge t > \infty \\ \downarrow \\ M_1 : 0 > t > \infty \\ \downarrow \\ U \rightarrow \begin{matrix} t \sim E_{Cell};\ TC: M_3 \rightarrow U \\ t \sim E_{Human};\ TC: U \rightarrow M_3 \end{matrix} \end{bmatrix}_{Time} & TC \rightarrow x_T
\end{array} \right\|_{\langle x_U | x_T \rangle}
\end{aligned}
$$

Starting with the Light-Matrix, the top left-hand matrix in (3.1.3), the first line from the top, $C_\infty:[Pr, Po, K, H]$, specifies the fundamental architecture of living cells. As explored previously in this chapter, Nucleic Acids are an emergence of Light's property of Knowledge, Polysaccharides are an emergence of Light's property of Power, Proteins are an emergence of Light's property of Presence, and Lipids are an emergence of Light's property of Harmony. The fundamental architecture of these aspects hence, is an emergence of the properties of Light at ∞.

Line 3 in the Light-Matrix, $C_K: [S_{Pr}, S_{Po}, S_K, S_H]$, elaborates the sets for Presence, Power, Knowledge, and Harmony, each containing multiple elements. For example, as explored previously in the section on Proteins, various elements derived from the four sets would define the behavior of Proteins and could be functions such as 'exist for service', 'to bring about perfection at the level of the cell', 'extreme diligence and perseverance', amongst others, hence collectively describing Proteins' way of being. Specifically, Line 5, $C_{N:}\, f(S_{Pr}\ x\ S_{Po}\ x\ S_K\ x\ S_H)$, suggests that unique seeds are created from a combination of such elements from all four sets, with a particular element leading, that in effect creates the distinctness possible at the level of cells.

Line 6, $(\downarrow R_{C_U} = f(R_{C_N}))$, specifies quantization between the layer where the seeds are formed, and the physical layer we are familiar with, and as explored in Chapter 3.5 and 3.6, will result in Line 7, C_U: $[P,V,M,C]$, hence changing the material fabric of existence. The possibilities represented by Lines 1 through 5 hence concretize through the quantization represented by Line 6 to become or enhance living cells with its subtle physical (related to Presence), vital (related to Power), mental (related to Knowledge), and connection (related to Harmony) aspects now existing in material reality typified by Light moving at c. Note that just as Line 6 represents a process of quantization relating the layer of reality created by Light traveling at c with the antecedent layers, so too Lines 2 and 4 as previously discussed, also represent quantization of a more subtle kind that ultimately plays a critical part in allowing the material fabric to express infinite diversity.

Typically it is the process as captured by the Space-Matrix that will determine if Line 6 is activated. Specifically patterns at the untransformed layer, U, will need to be overcome, as specified by the second-line from the bottom of the Space-Matrix: $(\uparrow > P_x)$. But as specified by the bottom-line of the Time-Matrix, reproduced below, it is only with the advent of the human-system that the automaticity of the action of meta-levels is reversed:

$$U \rightarrow \begin{matrix} t \sim E_{Cell}; TC: M_3 \rightarrow U \\ t \sim E_{Human}; TC: U \rightarrow M_3 \end{matrix}$$

Hence in the case of living cells, which in this emergence is a pre-human system, the fact that patterns do not need to be overcome means that quantization more or less happens automatically.

Nonetheless, and given this context it is useful to review
Equation 3.6.5, Quantization Effect of Organization on Material Fabric:

Impact on Materal Fabric

$$= \begin{bmatrix} \|[L][S][T]TC \rightarrow x_T\|_{\langle x_U|x_T\rangle} \\ \times \\ \left(\left|mod(Z_Q)\right|_{Y>U}\right) \\ \ni \\ Z_Q, Z \in \mathbb{U}\ (Space, Time, Energy, Gravity) \end{bmatrix}$$

Line 1 in the matrix is simply (3.6.1) the Simplified Light-Space-Time Emergence equation. To understand if quantization is going to occur, the organization resulting from the application of Line 1 is, where 'is' means 'multiplied' or $(\times)$, by a 'modulation', specified by (mod), of a fourfold space-time-energy-gravity quantization (Z_Q) kind, so long as Y>U, that is, the operative bases are relatively transformed and hence have an active influence greater than U. The fourfold quantization is specified by $(\ni Z_Q, Z \in U\ (Space, Time, Energy, Gravity))$ where $(\ni)$ the quantization being applied (Z) is each (U) of the elements of the set (Space, Time, Energy, Gravity).

But as just summarized in the Time-Matrix in (3.1.3) Y is by definition greater than U and hence quantization is automatic. In terms of living cells such quantization implies that wholeness becomes fully active through specific space, time, energy, and gravity quantization to create an holistic "ecosystem" with its own "living cell logic" as it were. The wholeness has now precipitated into the material fabric and is available to be consciously and unconsciously tapped into.

Note that while living cells as one of the earlier life manifestations of Light exists pervasively on earth and perhaps elsewhere, technologies that leverage its deeper or implicit aspects, such as the intent of various molecular plans that comprise the cells, will need the full operation of the Time-Matrix starting with the overcoming of patterns at U by human systems, in order to be fully leveraged. Such dynamics that in a sense 'evolve' the functioning of the living cell system will be reviewed in more detail in Section 10.

Chapter 6.2: Sensations, Urges, Feelings, Thoughts

As human beings we experience sensations, urges and desires and wills, feelings and emotions, and thought. These are key aspects of our being and becoming and critical aspects of how choice at both the individual and collective levels may be determined.

This chapter will go over these fundamental aspects of self.

Sensations

Sensations are those things we experience with our senses. We see things, hear things, and smell things, taste things, can touch things. This ability to enter into relationship with objects through sensation is nothing other than a result of the emergence of Light's property of Presence. We become present to Presence through the device of sensation. Sensation can be thought of as the means by which this property of Light – Presence - molds or ingrains itself in us as human beings. Its potentiality, all which is contained in this aspect of Light, becomes available to us through the power of sensation. Hence an equation, Equation 6.2.1, will generally represent the family of sensations. Some elements that it would comprise of may be 'tangible', 'take notice of', amongst others.

$$Sig_{sensation} = Xa + \overline{Yb_{0-n}} \quad where \left[\begin{array}{c} X \in [S_{System_{Pr}}] \\ Y \in [S_{System_{Pr}}, S_{System_{P}}, S_{System_{K}}, S_{System_{N}}] \\ a, b \; are \; integers; a > b \end{array} \right]$$

Eq 6.2.1: Sensation

There could also be equations for hearing, seeing, tasting, touching, and smelling.

But there is also a deeper experience of sensation that is possible. When we see things, for instance, what are we seeing? Is it just the surface rendering of the play of matter, or do we see that the fullness of Light is still there, with all its potentiality and possibility, in the smallest thing we look at? Do we see that the whole universe and more is present in all its fullness in the least thing that we easily ignore, or belittle, or loathe? When we touch things is it the seeming concreteness of the play of the particles or atoms or chains of molecules that we touch? Or is it the Love and Light and the vastness of all that IS that allows itself to be as a small corner that we touch so as to make infinity be felt by something so finite?

Such a deeper contact offered through sensation suggests a subset of (6.2.1) with secondary elements perhaps described as 'fullness of Light', 'contacting infinity', amongst others, thus also yielding an equation form (6.2.2):

$$Sig_{deeper-sensation} = Xa + \overline{Yb_{0-n}} \quad where \begin{bmatrix} X \in [S_{System_{P_T}}] \\ Y \in [S_{System_{P_T}}, S_{System_P}, S_{System_K}, S_{System_N}] \\ a, b \text{ are integers}; a > b \end{bmatrix}$$

Eq 6.2.2: Deeper Sensation

Urges, Desires & Wills

Urges and desires and wills are similarly a play of the emergence of Light's property of Power. In the mystery of focus, the vastness of Light has projected itself in us into an apparent smallness that is in reality everything that is. And this smallness is trying through urge and desire and will to connect viscerally or even intentionally to other smallnesses that similarly are nothing other than the fullness of Light projected into a small smorgasbord of selected function. So the urge or desire for food, or companionship, or of possession, or of climbing a peak, is nothing other than Light's compressed property of Power, trying to reach more of the fullness that it is through a fulfillment of the urge or desire or will that it masquerades as. Hence urges can be represented as Equation 6.2.3:

$$Sig_{urges} = Xa + \overline{Yb_{0-n}} \quad where \begin{bmatrix} X \in [S_{System_P}] \\ Y \in [S_{System_{P_T}}, S_{System_P}, S_{System_K}, S_{System_N}] \\ a, b \text{ are integers}; a > b \end{bmatrix}$$

Eq 6.2.3: Urges

Elements may be of the type of 'grasp', 'possess', 'deeply connect', amongst others.

Feelings & Emotions

Feelings and emotions are a play of the emergence of Light's property of Harmony or Nurturing. Its instrument is the Heart and it generates an array of emotions that are an indication or active radar of whether we are moving toward or away from a reality of harmony, whether based on our small self or some larger Self of Light. Gradually, by navigating with these emotions and feelings we can get to a state where we always feel positive emotions which basically means we have more truly entered into relationship with some larger continent of Light. An equation for feelings is as represented by Equation 6.2.4:

$$Sig_{feelings} = Xa + \overline{Yb_{0-n}} \ \ where \left[\begin{array}{c} X \in [S_{System_N}] \\ Y \in [S_{System_{P_T}}, S_{System_P}, S_{System_K}, S_{System_N}] \\ a, b \ are \ integers; a > b \end{array} \right]$$

Eq 6.2.4: Feelings

Thoughts

Thoughts are a play of the emergence of Light's property of Knowledge. Through the thought we can become greater or conceptualize things greater or begun to enter into relationship with some things other than our small self. Thought allows us to connect to more "othernesses" or even the oneness of the reality of Light.

$$Sig_{thoughts} = Xa + \overline{Yb_{0-n}} \ \ where \left[\begin{array}{c} X \in [S_{System_K}] \\ Y \in [S_{System_{P_T}}, S_{System_P}, S_{System_K}, S_{System_N}] \\ a, b \ are \ integers; a > b \end{array} \right]$$

Eq 6.2.5: Thoughts

So we can see that the very substance of our becoming – sensations, urges, wills, desires, feelings, emotions, thoughts – are nothing other than an emergence of the four properties of light of presence, power, knowledge, and harmony.

Light's Emergence as Fundamental Capacities of Self

As discussed previously the Light-Space-Time Emergence equation (3.1.3) being iterative, can be used to model emergence as it proceeds from simpler four-fold to more complex four-fold manifestations. Hence (3.1.3) has already been applied to suggest the emergence of the electromagnetic spectrum, quantum particles in general, bosons as a further instance of a particular kind of quantum particle, atoms, and living cells. Here it will be applied to suggest the emergence of fundamental capacities of self. But further, as implied by (3.6.5), the Quantization Effect of Organization on Material Fabric equation, any process of deeper organization, such as is responsible for the architecture and cohesiveness of capacities of self, has the possibility of altering the material fabric or fabric of existence so long as the bases involved are driven primarily by a meta-level.

As suggested by (3.1.3), reproduced below for convenience, the architecture and details of capacities of self can be seen to be the result of the application of the Light, Space, and Time matrices as will be elaborated:

$$Emergence_{light-space-time} = \left|\begin{matrix} \left[\begin{matrix} C_{\infty}: [Pr, Po, K, H] \\ (\downarrow R_{C_K} = f(R_{C_\infty})) \\ C_K: [S_{Pr}, S_{Po}, S_K, S_H] \\ (\downarrow R_{C_N} = f(R_{C_K})) \\ C_N: f(S_{Pr}\ x\ S_{Po}\ x\ S_K\ x\ S_H) \\ (\downarrow R_{C_U} = f(R_{C_N})) \\ C_U: [P, V, M, C] \end{matrix}\right]_{Light} & \left[\begin{matrix} M_3 \rightarrow System_X \\ (\uparrow F \rightarrow I) \\ M_2 \rightarrow S_{System\,X} \\ (\uparrow Sig \rightarrow F) \\ M_1 \rightarrow Sig_x \\ (\uparrow > P_x) \\ U \rightarrow x_U \end{matrix}\right]_{Space} \\ \left[\begin{matrix} & M_3 : -\infty \leq t \leq \infty \\ & \downarrow \\ & M_2 : 0 \geq t > \infty \\ & \downarrow \\ & M_1 : 0 > t > \infty \\ & \downarrow \\ U \rightarrow & \begin{matrix} t \sim E_{Cell}; \text{TC}: M_3 \rightarrow U \\ t \sim E_{Human}; \text{TC}: U \rightarrow M_3 \end{matrix} \end{matrix}\right]_{Time} & TC \rightarrow x_T \end{matrix}\right|_{\langle x_U | x_T \rangle}$$

Starting with the Light-Matrix, the top left-hand matrix in (3.1.3), the first line from the top, C_∞:$[Pr, Po, K, H]$, specifies the architecture of the fundamental capacities of self introduced in this chapter. Hence, Thoughts are an emergence of Light's property of Knowledge, Urges, Desires, and Wills are an emergence of Light's property of Power, Sensations are an emergence of Light's property of Presence, and Feelings and Emotions are an emergence of Light's property of Harmony. The fundamental architecture of these aspects hence, is an emergence of the properties of Light at ∞.

Line 3 in the Light-Matrix, C_K: $[S_{Pr}, S_{Po}, S_K, S_H]$, elaborates the sets for Presence, Power, Knowledge, and Harmony, each containing multiple elements. For example, as explored previously in the section on Sensations, various elements derived from the four sets would define the behavior of Sensations and could be functions such as 'tangible', 'take notice of', amongst others, hence collectively describing Sensations' way of being. Specifically, Line 5, C_N: $f(S_{Pr}\, x\, S_{Po}\, x\, S_K\, x\, S_H)$, suggests that unique seeds are created from a combination of such elements from all four sets, with a particular element leading, that in effect creates the distinctness possible at the level of fundamental capacities of self.

Line 6, $(\downarrow R_{C_U} = f(R_{C_N}))$, specifies quantization between the layer where the seeds are formed, and the physical layer we are familiar with, and as explored in Chapter 3.5 and 3.6, will result in Line 7, C_U: $[P, V, M, C]$, hence changing the material fabric of existence. The possibilities represented by Lines 1 through 5 hence concretize through the quantization represented by Line 6 to become or further enhance the capacities of self with its subtle physical (related to Presence), vital (related to Power), mental (related to Knowledge), and connection (related to Harmony) aspects now existing in material reality typified by Light moving at c. Note that just as Line 6 represents a process of quantization relating the layer of reality created by Light traveling at c with the antecedent layers, so too Lines 2 and 4 as previously discussed, also represent quantization of a more subtle kind that ultimately plays a critical part in allowing the material fabric to express infinite diversity.

Typically it is the process as captured by the Space-Matrix that will determine if Line 6 is activated. Specifically patterns at the untransformed layer, U, will need to be overcome, as specified by the second-line from the bottom of the Space-Matrix: $(\uparrow > P_x)$. But as specified by the bottom-line of the Time-Matrix, reproduced below, it is only with the advent of the human-system that the automaticity of the action of meta-levels is reversed:

$$U \rightarrow \begin{matrix} t \sim E_{Cell};TC: M_3 \rightarrow U \\ t \sim E_{Human};TC: U \rightarrow M_3 \end{matrix}$$

Hence in the case of sensations, which in this emergence is largely a post-human system, the fact that patterns do need to be overcome means that quantization requires effort to happen. Given this, it is useful to review Equation 3.6.5, Quantization Effect of Organization on Material Fabric:

Impact on Materal Fabric

$$= \begin{bmatrix} |[L][S][T]TC \rightarrow x_T|_{\langle x_U|x_T\rangle} \\ \times \\ \left(|mod(Z_Q)|_{Y>U}\right) \\ \ni \\ Z_Q, Z \in \mathbb{U}\ (Space, Time, Energy, Gravity) \end{bmatrix}$$

Line 1 in the matrix is simply (3.6.1) the Simplified Light-Space-Time Emergence equation. To understand if quantization is going to occur, the organization resulting from the application of Line 1 is, where 'is' means 'multiplied' or $(\times)$, by a 'modulation', specified by (mod), of a fourfold space-time-energy-gravity quantization (Z_Q) kind, so long as Y>U, that is, the operative bases are relatively transformed and hence have an active influence greater than U. The fourfold quantization is specified by $(\ni Z_Q, Z \in U\ (Space, Time, Energy, Gravity))$ where $(\ni)$ the quantization being applied (Z) is each (U) of the elements of the set (Space, Time, Energy, Gravity).

But as just summarized in the Time-Matrix in (3.1.3) Y is by definition not greater than U and hence quantization is not automatic.

In terms of sensations such quantization implies that increasing wholeness can become fully active through specific space, time, energy, and gravity quantization to create an holistic "ecosystem" with its own "sensation logic" as it were. The wholeness only then precipitates into the material fabric and is available to be consciously and unconsciously tapped into.

Taking the example of getting to deeper experiences of sensation as described previously in this chapter: "When we see things, for instance, what are we seeing? Is it just the surface rendering of the play of matter, or do we see that the fullness of Light is still there, with all its potentiality and possibility, in the smallest thing we look at? Do we see that the whole universe and more is present in all its fullness in the least thing that we easily ignore, or belittle, or loathe? When we touch things is it the seeming concreteness of the play of the particles or atoms or chains of molecules that we touch? Or is it the Love and Light and the vastness of all that IS that allows itself to be as a small corner that we touch so as to make infinity be felt by something so finite?"

As described, such a deeper contact offered through sensation suggests a subset of (6.2.1) with secondary elements perhaps described as 'fullness of Light', 'contacting infinity', amongst others, thus also yielding an equation form (6.2.2). But this implies a complete breaking of patterns associated with sensation such that Y>U and the quantization referred to in (3.6.5) can become operative to indeed change the fabric of material existence making such deeper type of sensation possible.

Chapter 6.3: Truer Individuality

As discussed in Chapter 1.3, at the core individuals can be thought of as projections of seeds formed in a vast continent of Light. So that core must always be there but it is often covered by surface dynamics of the physical, the vital, and the mental. These too are formed from properties of Light, but since the light has been separated from its source these movements and dynamics are incomplete in their nature.

It is easy for these dynamics and movements to completely occupy all individual processing power. This can happen to such an extent that the individual acting on the surface can become subject to these surface movements. But so long as there is the remembrance that even in separation and smallness - whether selfishness, myopia, fundamentalism, or any other ism - these movements are still light, then that can become the means by which the apparent chains around individuals can be loosened.

For in its essence these small movements are trying to extend into something other than what they are. By connecting the light in them with the larger Light this extension can yield something different than what is possible through any other means. And in the process the hold that these smaller movements have is diminished and one can begin to see or experience larger vistas of light.

Such a passage where one is always moving to larger vistas of light, more easily allows entry into a field of original seeds and it is so that each can begin to enter into conscious communion with truer individuality.

The truer individuality itself is some mathematical function of the four properties of Light - Harmony, Power, Knowledge, and Service, which in turn, are practically infinite sets of qualities related to one of the main properties of Light.

So it may be that one individual is primarily driven to connect people together, itself a variation of the property of Harmony. The individual may further want to do this by deeply understanding what makes these people each tick, itself a variation of the property of Knowledge. So the seed or the truer individuality of this person can be thought of as a mathematical function consisting of some element or property from the set of Harmony and some element or quality from the set of Knowledge, combined together with possibly different elements from the same or different sets, all with possibly different weights, but with the first weight of the need to connect people, being the strongest.

Such an individual may have an essential form as expressed in Equation 6.3.1, Harmony-Type Individual:

$$Sig_{Harmony-type} = Xa + \overline{Yb_{0-n}} \quad where \begin{bmatrix} X \in [S_{System_N}] \\ Y \in [S_{System_{Pr}}, S_{System_P}, S_{System_K}, S_{System_N}] \\ a, b \ are \ integers; a > b \end{bmatrix}$$

Eq 6.3.1: Harmony-Type Individual

But equally, and consistent with the emergent properties of Light there could be essential Service-Type, Power-Type, and Knowledge-Type individuals as well, with infinite variation in the precise nature of the core seed, as captured in Equations 6.3.2-4:

$$Sig_{Service-type} = Xa + \overline{Yb_{0-n}} \quad where \begin{bmatrix} X \in [S_{System_{Pr}}] \\ Y \in [S_{System_{Pr}}, S_{System_P}, S_{System_K}, S_{System_N}] \\ a, b \ are \ integers; a > b \end{bmatrix}$$

Eq 6.3.2: Service-Type Individual

$$Sig_{Power-type} = Xa + \overline{Yb_{0-n}} \quad where \begin{bmatrix} X \in [S_{System_P}] \\ Y \in [S_{System_{Pr}}, S_{System_P}, S_{System_K}, S_{System_N}] \\ a, b \ are \ integers; a > b \end{bmatrix}$$

Eq 6.3.3: Power-Type Individual

$$Sig_{Knowledge-type} = Xa + \overline{Yb_{0-n}} \quad where \begin{bmatrix} X \in [S_{System_K}] \\ Y \in [S_{System_{Pr}}, S_{System_P}, S_{System_K}, S_{System_N}] \\ a, b \ are \ integers; a > b \end{bmatrix}$$

Eq 6.3.4: Knowledge-Type Individual

Connecting with this seed or truer individuality, and then always keeping that reality alive means that the play of light is altered, so that the deeper light from or closer to the continent of Light is allowed to participate more actively in the practical everyday time-space world, that is itself a result of the slowing down of the speed of light. So now this separated world begins to be more connected to the continent of Light through truer individuality.

Truer individuality is allowing the separation between worlds or layers of light to dissolve and is causing this separated world to enter back into its nature of eternity or timelessness and spacelessness or infinity. Except that the individualities are now in concrete material form that then becomes a far more beautiful basis for continued projections and crystallizations from the continent of Light. But this can also be thought of as the result of four-fold quantization.

Light's Emergence as Truer Individuality

As discussed previously the Light-Space-Time Emergence equation (3.1.3) being iterative, can be used to model emergence as it proceeds from simpler four-fold to more complex four-fold manifestations. Hence (3.1.3) has already been applied to suggest the emergence of the electromagnetic spectrum, quantum particles in general, bosons as a further instance of a particular kind of quantum particle, atoms, living cells, and basic capacities of the self. Here it will be applied to suggest the emergence of truer individuality. But further, as implied by (3.6.5), the Quantization Effect of Organization on Material Fabric equation, any process of deeper organization, such as is responsible for the architecture and cohesiveness of truer individuality, has the possibility of altering the material fabric or fabric of existence so long as the bases involved are driven primarily by a meta-level.

As suggested by (3.1.3), reproduced below for convenience, the architecture and details of truer individuality can be seen to be the result of the application of the Light, Space, and Time matrices as will be elaborated:

$$Emergence_{light-space-time} = \left|\begin{matrix} \begin{bmatrix} C_{\infty}: [Pr, Po, K, H] \\ (\downarrow R_{C_K} = f(R_{C_{\infty}})) \\ C_K: [S_{Pr}, S_{Po}, S_K, S_H] \\ (\downarrow R_{C_N} = f(R_{C_K})) \\ C_{N:} f(S_{Pr} \, x \, S_{Po} \, x \, S_K \, x \, S_H) \\ (\downarrow R_{C_U} = f(R_{C_N})) \\ C_U: [P, V, M, C] \end{bmatrix}_{Light} & \begin{bmatrix} M_3 \rightarrow System_X \\ (\uparrow F \rightarrow I) \\ M_2 \rightarrow S_{System\, X} \\ (\uparrow Sig \rightarrow F) \\ M_1 \rightarrow Sig_x \\ (\uparrow > P_x) \\ U \rightarrow x_U \end{bmatrix}_{Space} \\ \begin{bmatrix} M_3 : -\infty \leq t \leq \infty \\ \downarrow \\ M_2 : 0 \geq \mathrm{t} > \infty \\ \downarrow \\ M_1 : 0 > \mathrm{t} > \infty \\ \downarrow \\ U \rightarrow \begin{matrix} \mathrm{t} \sim E_{Cell}; \mathrm{TC}: M_3 \rightarrow \mathrm{U} \\ \mathrm{t} \sim E_{Human}; \mathrm{TC}: U \rightarrow M_3 \end{matrix} \end{bmatrix}_{Time} & TC \rightarrow x_T \end{matrix}\right|_{(x_U | x_T)}$$

Starting with the Light-Matrix, the top left-hand matrix in (3.1.3), the first line from the top, C_{∞}:[Pr, Po, K, H], specifies the architecture of truer individuality as already discussed in this chapter. Hence, Knowledge-type individuals are an emergence of Light's property of Knowledge, Power-type individuals are an emergence of Light's property of Power, Service-type individuals are an emergence of Light's property of Presence, and Harmony-type individuals are an emergence of Light's property of Harmony. The fundamental architecture of these aspects hence, is an emergence of the properties of Light at ∞.

Line 3 in the Light-Matrix, C_K: [S_{Pr}, S_{Po}, S_K, S_H], elaborates the sets for Presence, Power, Knowledge, and Harmony, each containing multiple elements. For example, as just explored in the previous section, various elements derived from the four sets would define the architecture of Harmony-type individuals and could be functions such as 'driven to connect people together', amongst others, hence collectively describing Harmony-type individuals' way of being. Specifically, Line 5, $C_{N:} f(S_{Pr} \, x \, S_{Po} \, x \, S_K \, x \, S_H)$, suggests that unique seeds are created from a combination of such elements from all four sets, with a particular element leading, that in effect creates the distinctness possible at the level of individuals.

Line 6, $(\downarrow R_{C_U} = f(R_{C_N}))$, specifies quantization between the layer where the seeds are formed, and the physical layer we are familiar with, and as explored in Chapter 3.5 and 3.6, will result in Line 7, C_U: $[P,V,M,C]$, hence changing the material fabric of existence. The possibilities represented by Lines 1 through 5 hence concretize through the quantization represented by Line 6 to become or further enhance the unique individuality with its subtle physical (related to Presence), vital (related to Power), mental (related to Knowledge), and connection (related to Harmony) aspects now existing in material reality typified by Light moving at c. Note that just as Line 6 represents a process of quantization relating the layer of reality created by Light traveling at c with the antecedent layers, so too Lines 2 and 4 as previously discussed, also represent quantization of a more subtle kind that ultimately plays a critical part in allowing the material fabric to express infinite diversity.

Typically it is the process as captured by the Space-Matrix that will determine if Line 6 is activated. Specifically patterns at the untransformed layer, U, will need to be overcome, as specified by the second-line from the bottom of the Space-Matrix: $(\uparrow > P_x)$. But as specified by the bottom-line of the Time-Matrix, reproduced below, it is only with the advent of the human-system that the automaticity of the action of meta-levels is reversed:

$$U \rightarrow \begin{matrix} t \sim E_{Cell};TC: M_3 \rightarrow U \\ t \sim E_{Human};TC: U \rightarrow M_3 \end{matrix}$$

Hence in the case of truer individuality, which in this emergence is a post-human system, the fact that patterns do need to be overcome means that quantization requires effort to happen. Given this, it is useful to review Equation 3.6.5, Quantization Effect of Organization on Material Fabric:

Impact on Materal Fabric

$$= \begin{bmatrix} |[L][S][T]TC \rightarrow x_T|_{\langle x_U | x_T \rangle} \\ \times \\ \left(|mod(Z_Q)|_{Y>U} \right) \\ \ni \\ Z_Q, Z \in \mathbb{U} \; (Space, Time, Energy, Gravity) \end{bmatrix}$$

Line 1 in the matrix is simply (3.6.1) the Simplified Light-Space-Time Emergence equation. To understand if quantization is going to occur, the organization resulting from the application of Line 1 is, where 'is' means 'multiplied' or $(\times)$, by a 'modulation', specified by (mod), of a fourfold space-time-energy-gravity quantization (Z_Q) kind, so long as Y>U, that is, the operative bases are relatively transformed and hence have an active influence greater than U. The fourfold quantization is specified by $(\ni Z_Q, Z \in U\ (Space, Time, Energy, Gravity))$ where $(\ni)$ the quantization being applied (Z) is each (U) of the elements of the set (Space, Time, Energy, Gravity).

But as just summarized in the Time-Matrix in (3.1.3) Y is by definition not greater than U and hence quantization is not automatic. In terms of truer individuality such quantization implies that increasing wholeness can become fully active through specific space, time, energy, and gravity quantization to create an holistic "ecosystem" with its own "truer individuality logic" as it were. The wholeness only then precipitates into the material fabric and is available to be consciously and unconsciously tapped into.

SECTION 7: EMERGENCE OF COMPLEX ORGANIZATIONS

Having traced the emergence of the four properties of Light through the primary fourfold emergence of space-time-energy-gravity, the electromagnetic spectrum, through matter, through life, and even through human becoming as manifest in sensation, urges, feelings, and thoughts, and truer individuality we now turn our attention to study the phenomenon of emergence in complex organizations.

In Chapter 7.1 an equation for sustainable organization is derived. Chapter 7.2 examines the emergence of mega-organizations. Chapter 7.3 examines the emergence of sustainable global civilization.

Chapter 7.1: Sustainability of Organizations

We have traced the emergence of the four properties of Light through the primary fourfold emergence of space-time-energy-gravity, the electromagnetic spectrum, through matter, through life, and even through human becoming as manifest in sensation, urges, feelings, and thoughts, and truer individuality. At each of these levels it seems it is a balanced combination of all four of the properties that creates stability and makes that structure or organism or being sustainable. And this has to be since the four properties occur simultaneously in light and must be impelled to hold to that relationship even when projected from it. Hence it should be possible to deductively arrive at an equation for organizational sustainability given the four-fold emergence of properties of light through subsequent layers of manifestation.

To triangulate, though, we briefly trace the four-fold emergence, reinforce the significance of this and inductively arrive at an equation for sustainability that we would have arrived at deductively anyway.

Starting with the primary emergence of space-time-energy-gravity we see that space is defined as that in which seeds of knowledge are planted – hence, existing as a means to express Light's property of knowledge. Time is defined as the inevitability of the seeds of knowledge maturing into fullness, in spite of any opposition. Hence time is seen as an emergence of Light's property of power. Energy is seen as that which accumulates to create matter. Hence it is an emergence of Light's property of Presence. Gravity is that by which relationship between objects comes into being – hence an emergence of Light's property of harmony.

With the electromagnetic spectrum we see that the spectrum as a whole is defined or architected by the speed of light c, as an emergence of nurturing, the wavelength λ, as an emergence of knowledge, the frequency f, as an emergence of power, and mass-potential as an emergence of presence. These occur simultaneously and constitute the wholeness of the electromagnetic spectrum.

Similarly, at the level of the quantum it can be seen that the integrity and functioning of the atom also depends on the integration of all four properties of light. Hence quarks, an emergence of knowledge, leptons, an emergence of power, bosons, an emergence of harmony or nurturing, and the Higgs-boson, an emergence of presence, act together to create the structure and functionality of every single atom.

What this reinforces is that every single thing must have been created through some integration and balance of the four properties of light.

Going up the organizational scale to the next level of complexity, to the level of the atom, the same pattern is found again: every single atom belongs to one of four groups, and in unison these four groups orchestrate the set of combinations of known compounds.

Hence, Alkali Metals and Alkali Earth Metals configured by the s-Group, from the family of system-power, Metals, Metalloid, Non-Metals, Halogens, and Noble Gases configured by the p-Group, from the family of system-knowledge, Transition Metals configured by the d-Group, from the family of system-presence, and Lanthanides and Actinides configured by the f-Group, from the family of system-nurturing, act together to create the complex array of compounds that form the entire material from which the physical and subsequently even the organic world around us is constructed.

Going up the organizational scale to another level of complexity, to the level of the cell, the same pattern is found again: basically every single cell of every single living creature that has ever been studied by humankind also has a similar balance and integration of these four sets of forces acting together.

Hence, proteins, from the family of presence, nucleic acids, from the family of knowledge, polysaccharides, from the family of power, and lipids, from the family of nurturing, work together to create the balanced functioning of every living cell.

The hypothesis hence, is that even at larger scale, in considering the most innovative organization at the level of teams, corporations, or markets, a similar integration and balance of the four sets of properties or forces may yield the best results. Recent developments in sustainability investment models ranging from Socially Responsible Investing (Logue, 2008), the Global Reporting Initiative (Willis,

2003), the Dow Jones Sustainability Index (Hope & Fowler, 2007), the Principles of Responsible Investing (Harvard-Edu, 2014), all reinforce the concept of investment criteria becoming broader to be based not only on economic, but on environmental, social, and governance factors as well. Further evidence suggests that financial returns on such broad-based investment models continually beat financial returns on regular funds such as the S&P 500, for instance (Openshaw, 2015). In their study of major transitions in evolution Smith and Szathmary (Smith & Szathmary, 1995) chronicle the development of life toward increasing complexity as an application of successful collaboration and even co-evolution by which species evolve by changing together as system pressures increase. This hypothesis can be summarized by the following graph:

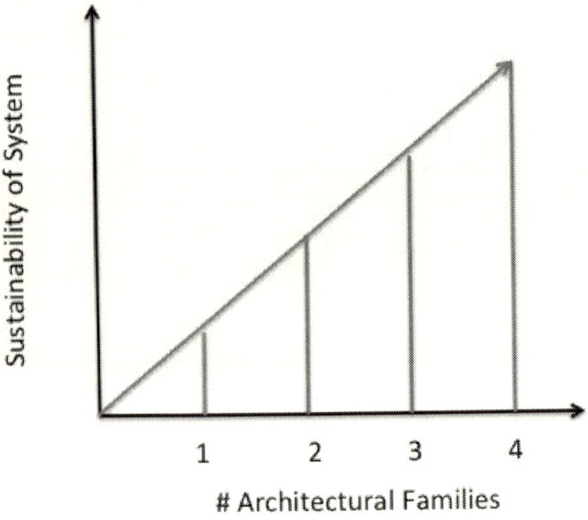

Figure 7.1.1 Sustainability of CAS

An equation for the sustainability of systems, $Sustainability_{Systems}$, where the interaction between the four families of forces is instrumental can be created. Hence, as in Equation 7.1.1:

$$Sustainability_{Systems} \propto Interaction\ (S_{System_{Pr}}, S_{System_{P}}, S_{System_{K}}, S_{System_{N}})$$

Eq 7.1.1: Sustainability of Systems

Chapter 7.2: Creation of a Stable Mega-Organization

As indicated by Equation 7.1.1, a sustainable organization is the result of two dynamics: maturity along each of the architectural force dimensions, and a robust interaction between all for sets of forces.

Consider the example of Silicon Valley.

While Silicon Valley may be primarily a commercial organization, yet we begin to glimpse something of what may become possible when the four emergent forces of light can begin to interact with one another in free fashion. That is, without the motive of doing everything for the sake of generating money only. In the case of Silicon Valley we see that the climate and beauty of the Bay Area, thereby likely representing the impetus of nurturing and harmony, began to attract many talented people into the vicinity. Over time the talent pool became progressively diversified. Educational institutes, such as Stanford University and University of Berkeley cropped up and became centers of cutting-edge research, representing the impetus of knowledge. The Armed Forces were attracted to the area for the same reason, and came with their huge requirements for research for research's sake, and their funds in support of the area, representing the impetus of power. Graduates from the universities started companies that began to in turn support the universities with funds. Talent moved around from company to company like people from department to department. Thus we see that even when individual companies failed, Silicon Valley functioned as a larger organization and was able to retain the talent in the area, was able further, to buffer the shocks to some extent, hence preparing the ground for future waves of innovation. This reality of becoming a container for dynamic experimentation perhaps represents the impetus of service in that the container is serving the dynamic components existing within it.

Pragmatically speaking this dynamic of functioning as a larger mega-organization meant that some level of commercial-immunity began to develop, so that people losing their jobs was not necessarily considered as that stressful an event. There was a freeing, thus, from the purely commercial element. If the various powers of beauty, aesthetics, knowledge, power in the form of money, plus the myriad streams of talent – from engineers, scientists, managers, lawyers etc., did not exist in such close proximity, and further, if the various professionals could not thus support each other through their continued informal meetings, through the urge to create the next wave of innovation, through the urge to pursue progress for the sake of progress, then they would have left for other areas and the phenomenon of Silicon Valley would never have been.

Usually modern day organizations cannot provide these various buffers and opportunities for interaction that Silicon Valley provides, and hence when hit with adversity, more often than not, simply crumble. This is perhaps due to the fact that it is always only one-dimension that drives them, and when the health of that is threatened or falls, the organization resorts to tactics that will ensure that that dimension looks good at any cost, in the bargain sacrificing the development of the other dimensions, and therefore its longer term health.

The four-forces led organization will be something like a community. Having attained freedom from an exaggerated commercial impetus, people will 'live' their 'jobs' because it is what fulfills them. Such a freedom is what will allow the four primary powers to manifest to greater and greater degree within them and their environment. Seeing thus, their ability to become centers of knowledge and wisdom, or mutuality and harmony, or power and leadership and energy, or perfection and service, or some unique combination of these primary forces, so increase, a sense of satisfaction with life will more easily accompany all that they continue to do. Under the freer flow of these powers their uniqueness will be refined and flourish, and correspondingly, so too will the uniqueness of their respective organizations.

An organization may be political and hence be primarily driven by the power of leadership and courage, or it may be social and hence be primarily driven by the power of service and perfection, or it may be commercial and hence be primarily driven by the power of mutuality and harmony, or it may be research-oriented and academic and hence be primarily driven by the powers of knowledge and wisdom, but always each of the other powers will also be behind it, fulfilling and completing its primary urge. Thus the dynamic of unique primary powers being supported by an increasing plethora of secondary powers is represented by the following four organizational equations, (7.2.1 – 4) and illustrates how the four powers of light emerge even in complex human-based collectivities:

$$Sig_{Political} = Xa + \overline{Yb_{0-n}} \quad where \begin{bmatrix} X \in [S_{System_P}] \\ Y \in [S_{System_{P_r}}, S_{System_P}, S_{System_K}, S_{System_N}] \\ a, b \text{ are integers}; a > b \end{bmatrix}$$

Eq 7.2.1: Political Organization

$$Sig_{Social} = Xa + \overline{Yb_{0-n}} \quad where \begin{bmatrix} X \in [S_{System_{P_r}}] \\ Y \in [S_{System_{P_r}}, S_{System_P}, S_{System_K}, S_{System_N}] \\ a, b \text{ are integers}; a > b \end{bmatrix}$$

Eq 7.2.2: Social Organization

$$Sig_{Academic} = Xa + \overline{Yb_{0-n}} \quad where \begin{bmatrix} X \in [S_{System_K}] \\ Y \in [S_{System_{P_r}}, S_{System_P}, S_{System_K}, S_{System_N}] \\ a, b \text{ are integers}; a > b \end{bmatrix}$$

$$Sig_{Commercial} = Xa + \overline{Yb_{0-n}} \quad where \begin{bmatrix} X \in [S_{System_N}] \\ Y \in [S_{System_{P_r}}, S_{System_P}, S_{System_K}, S_{System_N}] \\ a, b \text{ are integers}; a > b \end{bmatrix}$$

Eq 7.2.3: Commercial Organization

$$Sig_{Academic} = Xa + \overline{Yb_{0-n}} \quad where \begin{bmatrix} X \in [S_{System_K}] \\ Y \in [S_{System_{P_r}}, S_{System_P}, S_{System_K}, S_{System_N}] \\ a, b \text{ are integers}; a > b \end{bmatrix}$$

Eq 7.2.4: Academic Organization

Leveraging Mathematical Operators to Promote Organizational and Ecosystem Evolution

Note too that the host of mathematical operators derived deductively in Chapter 2.8, and represented by (2.8.1-17) can also be leveraged to support the emergence of the deeper powers represented by (7.2.1-4) and further enhance the overall ecosystem to operate in a more transformative way. These operators have to be considered in context to the Generalized Equation for Innovation, Equation 2.6.6, derived in Chapter 2.6. This equation suggests that any system has implicit in it the urge to transform the untransformed layer, U, by opening to the influence of the meta-layers, M_1, M_2, and M_3. In so doing the very sources of innovation are altered and the visible characteristics of systems are transformed by the action of these sources of innovation. Several sets were suggested that explore these sources of innovation - $S_{System_{Pr}}$, S_{System_P}, S_{System_K}, and S_{System_N}, and the resultant characteristics of systems - $Physical_T$, $Vital_T$, $Mental_T$, and $Integral_T$.

Even the number of "unique" people will continue to increase, become guiding lights so that the community will spontaneously begin to move in holistic directions consistent with the urge of people, of the sub-organizations within the community, of the community itself, and of the larger system of which the community or mega-organization is a part.

Even perhaps, something that may be behind the four archetypal powers, something that may be behind the building of all uniqueness – individual and organizational, more of the fullness of light and love, that is, may be compelled to come forward, seeing how developed its means of expression and action have become.

Light's Emergence as Mega-Organization

As discussed previously the Light-Space-Time Emergence equation (3.1.3) being iterative, can be used to model emergence as it proceeds from simpler four-fold to more complex four-fold manifestations. Hence (3.1.3) has already been applied to suggest the emergence of the electromagnetic spectrum, quantum particles in general, bosons as a further instance of a particular kind of quantum particle, atoms, living cells, basic capacities of the self, and truer individuality. Here it will be applied to suggest the emergence of mega-organization. But further, as implied by (3.6.5), the Quantization Effect of Organization on Material Fabric equation, any process of deeper organization, such as is responsible for the architecture and cohesiveness of mega-organizations, has the possibility of altering the material fabric or fabric of existence so long as the bases involved are driven primarily by a meta-level.

As suggested by (3.1.3), reproduced below for convenience, the architecture and details of the mega-organization can be seen to be the result of the application of the Light, Space, and Time matrices as will be elaborated:

$$
\begin{aligned}
&Emergence_{light-space-time} \\
&= \left|
\begin{matrix}
\begin{bmatrix}
C_{\infty}: [Pr, Po, K, H] \\
(\downarrow R_{C_K} = f(R_{C_{\infty}})) \\
C_K: [S_{Pr}, S_{Po}, S_K, S_H] \\
(\downarrow R_{C_N} = f(R_{C_K})) \\
C_N: f(S_{Pr}\ x\ S_{Po}\ x\ S_K\ x\ S_H) \\
(\downarrow R_{C_U} = f(R_{C_N})) \\
C_U: [P, V, M, C]
\end{bmatrix}_{Light}
&
\begin{bmatrix}
M_3 \rightarrow System_X \\
(\uparrow F \rightarrow I) \\
M_2 \rightarrow S_{System X} \\
(\uparrow Sig \rightarrow F) \\
M_1 \rightarrow Sig_x \\
(\uparrow > P_x) \\
U \rightarrow x_U
\end{bmatrix}_{Space}
\\
\begin{bmatrix}
M_3 : -\infty \leq t \leq \infty \\
\downarrow \\
M_2 : 0 \geq t > \infty \\
\downarrow \\
M_1 : 0 > t > \infty \\
\downarrow \\
U \rightarrow \begin{matrix} t \sim E_{Cell};\ TC: M_3 \rightarrow U \\ t \sim E_{Human};\ TC: U \rightarrow M_3 \end{matrix}
\end{bmatrix}_{Time}
&
TC \rightarrow x_T
\end{matrix}
\right|_{\langle x_U | x_T \rangle}
\end{aligned}
$$

Starting with the Light-Matrix, the top left-hand matrix in (3.1.3), the first line from the top, $C_{\infty}:[Pr, Po, K, H]$, specifies the architecture of mega-organization as already discussed in this and the previous chapter. Hence, Academic organizations are an emergence of Light's property of Knowledge, Political organizations are an emergence of Light's property of Power, Social organizations are an emergence of Light's property of Presence, and Commercial organizations can be an emergence of Light's property of Harmony. The fundamental architecture of these aspects hence, is an emergence of the properties of Light at ∞.

Line 3 in the Light-Matrix, C_K: $[S_{Pr}, S_{Po}, S_K, S_H]$, elaborates the sets for Presence, Power, Knowledge, and Harmony, each containing multiple elements. For example, as just explored in the previous section, various elements derived from the four sets would define the architecture of a Silicon Valley type mega-commercial organization and could be functions such as 'commercial immunity', 'power of aesthetics', 'ease of informal meetings', amongst others, hence collectively describing mega-commercial organization's way of being. Specifically, Line 5, $C_{N:} f(S_{Pr}\, x\, S_{Po}\, x\, S_K\, x\, S_H)$, suggests that unique seeds are created from a combination of such elements from all four sets, with a particular element leading, that in effect creates the distinctness possible at the level of mega-organization.

Line 6, $(\downarrow R_{C_U} = f(R_{C_N}))$, specifies quantization between the layer where the seeds are formed, and the physical layer we are familiar with, and as explored in Chapter 3.5 and 3.6, will result in Line 7, C_U: $[P,V,M,C]$, hence changing the material fabric of existence. The possibilities represented by Lines 1 through 5 hence concretize through the quantization represented by Line 6 to become or further enhance the mega-organization with its subtle physical (related to Presence), vital (related to Power), mental (related to Knowledge), and connection (related to Harmony) aspects now existing in material reality typified by Light moving at c. Note that just as Line 6 represents a process of quantization relating the layer of reality created by Light traveling at c with the antecedent layers, so too Lines 2 and 4 as previously discussed, also represent quantization of a more subtle kind that ultimately plays a critical part in allowing the material fabric to express infinite diversity.

Typically it is the process as captured by the Space-Matrix that will determine if Line 6 is activated. Specifically patterns at the untransformed layer, U, will need to be overcome, as specified by the second-line from the bottom of the Space-Matrix: $(\uparrow > P_x)$. But as specified by the bottom-line of the Time-Matrix, reproduced below, it is only with the advent of the human-system that the automaticity of the action of meta-levels is reversed:

$$U \rightarrow \begin{matrix} t \sim E_{Cell};TC\text{: } M_3 \rightarrow U \\ t \sim E_{Human};TC\text{: } U \rightarrow M_3 \end{matrix}$$

Hence in the case of mega-organizations, which in this emergence is a post-human system, the fact that patterns do need to be overcome means that quantization requires effort to happen. Given this, it is useful to review Equation 3.6.5, Quantization Effect of Organization on Material Fabric:

$$\textit{Impact on Materal Fabric} = \left[\begin{array}{c} |[L][S][T]TC \rightarrow x_T|_{\langle x_U|x_T\rangle} \\ \times \\ \left(|mod(Z_Q)|_{Y>U}\right) \\ \ni \\ Z_Q, Z \in \mathbb{U}\ (Space, Time, Energy, Gravity) \end{array}\right]$$

Line 1 in the matrix is simply (3.6.1) the Simplified Light-Space-Time Emergence equation. To understand if quantization is going to occur, the organization resulting from the application of Line 1 is, where 'is' means 'multiplied' or $(\times)$, by a 'modulation', specified by (mod), of a fourfold space-time-energy-gravity quantization (Z_Q) kind, so long as Y>U, that is, the operative bases are relatively transformed and hence have an active influence greater than U. The fourfold quantization is specified by $(\ni Z_Q, Z \in U\ (Space, Time, Energy, Gravity))$ where $(\ni)$ the quantization being applied (Z) is each (U) of the elements of the set (Space, Time, Energy, Gravity).

But as just summarized in the Time Matrix in (3.1.3) Y is by definition not greater than U and hence quantization is not automatic. In terms of mega-organizations such quantization implies that increasing wholeness can become fully active through specific space, time, energy, and gravity quantization to create an holistic "ecosystem" with its own "mega-organizations logic" as it were. The wholeness only then precipitates into the material fabric and is available to be consciously and unconsciously tapped into.

Chapter 7.3: Creation of Sustainable Global Civilization

A brief look at history will reinforce the idea that it is typically maturity along multiple and distinct dimensions, represented by the four properties of light, and their rich interaction that allows civilizations to endure.

Thus, those civilizations that have endured typically have a balance of all four families (Sri Aurobindo, 1971). Civilizations that have become extinct typically have had a focus on few drivers of innovation. Jared Diamond proposes five interconnected causes of collapse that may reinforce each other: non-sustainable exploitation of resources, climate changes, diminishing support from friendly societies, hostile neighbors, and inappropriate attitudes for change (Diamond, 2005). But these five sources may also be thought of as symptoms that arise due to the failure to adopt the catholicity of the sources of innovation emanating from each of the four sets or families. Further, the historian Toynbee suggested that societies decay because of their over-reliance on structures that helped them solve old problems (Toynbee, 1961). It can be interpreted that being thus biased they are unable to adopt the catholicity of the sources of innovation emanating from each of the four sets of families.

Approaching Civilization from a big-picture, global basis though, the sustainability of humankind will be ensured by a balance of development amongst the four sets of forces. This means that countries and global regions must be unique and in such a way that their primary emergence is distributed amongst all four sets of forces. Further, and based on this uniqueness, there must be an open and healthy interaction amongst these centers of uniqueness.

As an illustration, the uniqueness of representative countries in equation form may be as follows.

Example of Knowledge: India

Historically at least India, for example, in its essence, may be thought of as having an exceptional capacity for penetrating behind the surface, and further of meaningfully synthesizing many streams of development. Its primary power may thus be from the family of knowledge, with a strong secondary driver being its ability to create living harmonies. The equation for India will then likely be represented by Equation 7.3.1:

$$Sig_{India} = Xa + \overline{Yb_{0-n}} \quad where \begin{bmatrix} X \in [S_{System_K}] \\ Y \in [S_{System_{Pr}}, S_{System_P}, S_{System_K}, S_{System_N}] \\ a, b \text{ are integers}; a > b \end{bmatrix}$$

Eq 7.3.1: India (Knowledge Family)

Example of Power: Japan

Japan, in its essence, may be thought of as having a strong and noble warrior nature, along with a refined sense of aesthetics, amongst other qualities. Its equation, Equation 7.3.2 would be of the form:

$$Sig_{Japon} = Xa + \overline{Yb_{0-n}} \quad where \begin{bmatrix} X \in [S_{System_P}] \\ Y \in [S_{System_{Pr}}, S_{System_P}, S_{System_K}, S_{System_N}] \\ a, b \text{ are integers}; a > b \end{bmatrix}$$

Eq 7.3.2: Japan (Power Family)

Example of Nurturing: UK

UK, in its essence, may be thought of as having a strong ability to create practical, materialistic harmonies, resulting in such things as working parliaments and advanced democracy, for example. Its equation, Equation 7.3.3, would be of the form:

$$Sig_{UK} = Xa + \overline{Yb_{0-n}} \quad where \begin{bmatrix} X \in [S_{System_N}] \\ Y \in [S_{System_{Pr}}, S_{System_P}, S_{System_K}, S_{System_N}] \\ a, b \text{ are integers}; a > b \end{bmatrix}$$

Eq 7.3.3: UK (Nurturing Family)

Example of Service: Thailand

Thailand, in its essence, may be characterized by an exceptional sense of hospitality and sweet service, with an attention to detail in the practical arrangement of things. Its equation, Equation 7.3.4, would be of the form:

$$Sig_{Thailand} =$$

$$Xa + \overline{Yb_{0-n}} \;\; where \begin{bmatrix} X \in [S_{System_N}] \\ Y \in [S_{System_{Pr}}, S_{System_P}, S_{System_K}, S_{System_N}] \\ a, b \; are \; integers; a > b \end{bmatrix}$$

Eq 7.3.4: Thailand (Service Family)

Emergence of Uniqueness

Similarly every nation on earth will have a uniqueness that can be represented by equations belonging to one of the four families.

As discussed in Chapter 2.5 on Emergence of Uniqueness, for uniqueness to emerge and mature is a process. The process is represented by Equation 2.5.1 reproduced here for convenience:

$$Sig_E = X \begin{vmatrix} C{:}Sig * mod\,(\int = 1) \\ F{:}Sig\; mod\,(c) \\ I{:}\; Sig\; mod\,(\int G, \bar{e}, \pi) \\ M{:}Sig * mod\,(G) \\ V{:}Sig * mod\,(e) \\ P{:}Sig * mod\,(\pi) \end{vmatrix}$$

Equations 7.3.1 – 4 represent example of the essence of uniqueness. For these to become living practicalities requires work at many different levels within a nation. If a nation has not really done the work to animate itself with its uniqueness then it may exist in the physical or P-state, or vital of V-state, or mental or M-state, which by definition lack sufficient maturity, and interaction with other nations is then going to be compromised. Moving to the integral or I-state, will allow a nation to at least not be locked in to a point of view. A consistent I-state practiced by each nation then is the minimum requirement for a sustainable global civilization. But a consistent I-state also means that restricting patterns are always being broken and that the quantization effect of organization on the material fabric is actively in play. In other words, separations of light are continually uniting with the larger continent of Light and changing the very nature of possibility on larger and larger scale. The next sub-section examines such a quantization effect in more detail.

Leveraging Mathematical Operators to Promote Organizational and Global Ecosystem Evolution

Note too that the host of mathematical operators derived deductively in Chapter 2.8, and represented by (2.8.1-17) can also be leveraged to support the emergence of the deeper powers represented by (7.3.1-4) and further enhance the overall global ecosystem to operate in a more transformative way. These operators have to be considered in context to the Generalized Equation for Innovation, Equation 2.6.6, derived in Chapter 2.6. This equation suggests that any system has implicit in it the urge to transform the untransformed layer, U, by opening to the influence of the meta-layers, M_1, M_2, and M_3. In so doing the very sources of innovation are altered and the visible characteristics of systems are transformed by the action of these sources of innovation. Several sets were suggested that explore these sources of innovation - $S_{System_{Pr}}$, S_{System_P}, S_{System_K}, and S_{System_N}, and the resultant characteristics of systems - $Physical_T$, $Vital_T$, $Mental_T$, and $Integral_T$.

Light's Emergence as Sustainable Global Civilization

As discussed in previous chapters the Light-Space-Time Emergence equation (3.1.3) being iterative, can be used to model emergence as it proceeds from simpler four-fold to more complex four-fold manifestations. Hence (3.1.3) has already been applied to suggest the emergence of the electromagnetic spectrum, quantum particles in general, bosons as a further instance of a particular kind of quantum particle, atoms, living cells, basic capacities of the self, truer individuality, and mega-organizations. Here it will be applied to suggest the emergence of sustainable global civilization. But further, as implied by (3.6.5), the Quantization Effect of Organization on Material Fabric equation, any process of deeper organization, such as is responsible for the architecture and cohesiveness of sustainable global civilization, has the possibility of altering the material fabric or fabric of existence so long as the bases involved are driven primarily by a meta-level.

As suggested by (3.1.3), reproduced below for convenience, the architecture and details of sustainable global civilization can be seen to be the result of the application of the Light, Space, and Time matrices as will be elaborated:

$$Emergence_{light-space-time}$$

$$= \left\| \begin{matrix} \begin{bmatrix} C_\infty: [Pr, Po, K, H] \\ (\downarrow R_{C_K} = f(R_{C_\infty})) \\ C_K: [S_{Pr}, S_{Po}, S_K, S_H] \\ (\downarrow R_{C_N} = f(R_{C_K})) \\ C_N: f(S_{Pr}\ x\ S_{Po}\ x\ S_K\ x\ S_H) \\ (\downarrow R_{C_U} = f(R_{C_N})) \\ C_U: [P, V, M, C] \end{bmatrix}_{Light} & \begin{bmatrix} M_3 \rightarrow System_X \\ (\uparrow F \rightarrow I) \\ M_2 \rightarrow S_{System_X} \\ (\uparrow Sig \rightarrow F) \\ M_1 \rightarrow Sig_x \\ (\uparrow > P_x) \\ U \rightarrow x_U \end{bmatrix}_{Space} \\ \begin{bmatrix} M_3 : -\infty \leq t \leq \infty \\ \downarrow \\ M_2 : 0 \geq t > \infty \\ \downarrow \\ M_1 : 0 > t > \infty \\ \downarrow \\ U \rightarrow \begin{matrix} t \sim E_{Cell}; \text{TC}: M_3 \rightarrow \text{U} \\ t \sim E_{Human}; \text{TC}: U \rightarrow M_3 \end{matrix} \end{bmatrix}_{Time} & TC \rightarrow x_T \end{matrix} \right\|_{\langle x_U | x_T \rangle}$$

Starting with the Light-Matrix, the top left-hand matrix in (3.1.3), the first line from the top, $C_\infty:[Pr, Po, K, H]$, specifies the architecture of sustainable global civilization as already discussed in this chapter and previous chapters. The key to such a civilization is pointed to by the previously derived equation (7.1.1), reproduced here for convenience:

$$Sustainability_{Systems} \propto Interaction\ (S_{System_{Pr}}, S_{System_P}, S_{System_K}, S_{System_N})$$

This requires maturity by large organized parts of the world, whether nations or regional blocs. The maturity is such that the fourfold properties of Light are adequately expressed. Hence in the example of nations illustrated earlier in this chapter, a country like India is in its deeper essence an emergence of Light's property of Knowledge, a country like Japan is in its deeper essence an emergence of Light's property of Power, a country like Thailand is in its deeper essence an emergence of Light's property of Presence, and a country like UK is in in its deeper essence an emergence of Light's property of Harmony. The fundamental architecture of the combination of such emergences, which as per (7.1.1) is required for sustainability, is hence an emergence of the properties of Light at ∞.

Line 3 in the Light-Matrix, $C_K\colon [S_{Pr},S_{Po},S_K,S_H]$, elaborates the sets for Presence, Power, Knowledge, and Harmony, each containing multiple elements. For example, as just explored in the previous section, various elements derived from the four sets would define the architecture of a knowledge-essence country like India and could be functions such as 'exceptional capacity for penetrating behind the surface', 'meaningfully synthesizing many streams of development', amongst others, hence collectively describing different aspects of knowledge-essence country's way of being. Specifically, Line 5, $C_{N:} f(S_{Pr}\, x\, S_{Po}\, x\, S_K\, x\, S_H)$, suggests that unique seeds are created from a combination of such elements from all four sets, with a particular element leading, that in effect creates the distinctness possible at the level of sustainable global civilization.

Line 6, $(\downarrow R_{C_U} = f(R_{C_N}))$, specifies quantization between the layer where the seeds are formed, and the physical layer we are familiar with, and as explored in Chapter 3.5 and 3.6, will result in Line 7, $C_U\colon [P,V,M,C]$, hence changing the material fabric of existence. The possibilities represented by Lines 1 through 5 hence concretize through the quantization represented by Line 6 to enhance sustainable global civilization with subtle physical (related to Presence), vital (related to Power), mental (related to Knowledge), and connection (related to Harmony) aspects now existing in material reality typified by Light moving at c. Note that just as Line 6 represents a process of quantization relating the layer of reality created by Light traveling at c with the antecedent layers, so too Lines 2 and 4 as previously discussed, also represent quantization of a more subtle kind that ultimately plays a critical part in allowing the material fabric to express infinite diversity.

Typically it is the process as captured by the Space-Matrix that will determine if Line 6 is activated. Specifically patterns at the untransformed layer, U, will need to be overcome, as specified by the second-line from the bottom of the Space-Matrix: $(\uparrow > P_x)$. But as specified by the bottom-line of the Time-Matrix, reproduced below, it is only with the advent of the human-system that the automaticity of the action of meta-levels is reversed:

$$U \to \begin{matrix} t \sim E_{Cell}; TC: M_3 \to U \\ t \sim E_{Human}; TC: U \to M_3 \end{matrix}$$

Hence in the case of sustainable global civilization, which in this emergence is a post-human system, the fact that patterns do need to be overcome means that quantization requires effort to happen. Given this, it is useful to review Equation 3.6.5, Quantization Effect of Organization on Material Fabric:

$$\textit{Impact on Materal Fabric} = \begin{bmatrix} |[L][S][T]TC \to x_T|_{\langle x_U | x_T \rangle} \\ \times \\ \left(|mod(Z_Q)|_{Y>U}\right) \\ \ni \\ Z_Q, Z \in \mathbb{U}\ (Space, Time, Energy, Gravity) \end{bmatrix}$$

Line 1 in the matrix is simply (3.6.1) the Simplified Light-Space-Time Emergence equation. To understand if quantization is going to occur, the organization resulting from the application of Line 1 is, where 'is' means 'multiplied' or $(\times)$, by a 'modulation', specified by (mod), of a fourfold space-time-energy-gravity quantization (Z_Q) kind, so long as Y>U, that is, the operative bases are relatively transformed and hence have an active influence greater than U. The fourfold quantization is specified by $(\ni Z_Q, Z \in U\ (Space, Time, Energy, Gravity))$ where $(\ni)$ the quantization being applied (Z) is each (U) of the elements of the set (Space, Time, Energy, Gravity).

But as just summarized in the Time-Matrix in (3.1.3) Y is by definition not greater than U and hence quantization is not automatic.

In terms of sustainable global civilization such quantization implies that increasing wholeness can become fully active through specific space, time, energy, and gravity quantization to create an holistic "ecosystem" with its own "sustainable global civilizastion logic" as it were. The wholeness only then precipitates into the material fabric and is available to be consciously and unconsciously tapped into.

SECTION 8: Reflection on Love & Light

This section seeks to tie together in a sheaf of love and light, the emergence of the properties of light from the early space-time-energy-gravity form to its current form of civilization.

Chapter 8.1: Love

In our ongoing exploration of Light we have discovered how it's slowing done could have created the whole universe through the Big Bang. We have also seen how the nature of realities that may exist is also dependent on the way light moves. So it can be seen that there are these incredible, existential mysteries in Light.

Here we are going to explore one more mystery of Light – that of Love.

Let us consider again each of the layers of pre-matter, matter, life, and complex organization we just explored. We can see that four properties of Light – Power, Knowledge, Harmony, and Service – have structured these layers. But in each of these cases it is not one of these, but all four working together that creates the foundation of the layer.

As the apparent first fourfold emergence a wholeness is apparent in space-time-energy-gravity that in a sense cerates the moorings of the universe. These exist together, as one, and it appears that it is only their combined action that sets the stage for all subsequent emergences. But what else is such a combined action that creates the floor and foundation of the universe, if not Love?

At the electromagnetic field level there is a sustainable wholeness that also comes into place because the possible archetypes embedded in the electromagnetic spectrum – the knowledge aspect – exists as a range of frequencies that allows different energies to come into being – the power aspect – and that can create a range of particles – the presence aspect – that all travel at a speed that allows what is created to interact – the harmony aspect. Without the participation of each of these aspects subsequent layers of matter could not emerge. But such complete interaction of different aspects working together to create a whole different possibility is nothing other than Love.

So we see so far that the properties of Light that project themselves out for the purpose of creation are bound together in an external wholeness by Love, which has too to be implicit or contained in Light itself.

But exploring further, at the quantum level, it can similarly be said that quarks or leptons or bosons or the Higgs-boson by themselves would not have been able to create the foundation for the next layer of complexity in Light's journey. If they acted independently there would have only been a flurry of endless flows. It is only the combined action of the four held together in a tight embrace that is able to create the reality of the atom. But such a tight embrace of distinct actors or personalities that causes cohesive and unified action is nothing other than Love.

Hence it is Love that pulls together the possibilities projected by Light to create a unity that must exist for any next layer of complexity to emerge.

Continuing with the exploration of the layer of quantum particles, it is the coordinated action of bosons that allows the richness, the possibilities, the dynamism of quantum particle interaction, and all the myriad results of this to come into being. Gluons allow the nucleus to form and therefore allow the sustained codification of what an element will mean in the scheme of things to come into being. W and Z bosons allow the release of energy to be recycled or used for other purposes of interaction, through decay of heavier quarks and leptons. Photons as the carrier particle of the electromagnetic spectrum allow this field to be ever-present in this reality. Hypothetical gravitons would allow built-up objects to enter into a space-time relationship with each other: it bounds objects to behave in a unified way in a specified space and for a specified time. But all this holding together, and releasing, and recycling, and relating, this cohesiveness of the action of the force-carriers or bosons is only another instance of the action of Love. Any of these forces acting by themselves would not be able to create the foundation on which subsequent emergences are built.

Similarly, the complexity of molecules resulting in the diversity of possibility is only so because such a molecular foundation is built by Love: it is Love that causes a sustainable molecular-foundation constructed from Power, Knowledge, Harmony, and Service type atoms to come into being. A universe created by one type of atom only would quickly be crushed under the weight of an ever more unsustainable imbalance.

Cells themselves require a balance of lipids, polysaccharides, proteins, and nucleic acids to meaningfully sustain themselves. Such balance is the sign that Love has left its stamp.

So where there was only Light, Love too was. Where Light has projected itself into ever-more crystallized forms of matter and life, there too Love must be. This is so because Light in its wholeness is inseparable. The properties of Light when projected therefore, still have to act as though one because that is the innate law of their being. This continued action of oneness in crystal form is what Love is, and it acts so that the properties of Knowledge, Power, Harmony, and Service can more fully manifest what Light really is.

But then as individuals come into being we see again that sensations, urges and wills and desires, emotions and feelings, and thoughts are nothing other than Light's emergences of Presence, Power, Harmony, and Knowledge. These all exist in each individual and allow the individual to be and become more meaningfully, also in an act of Grace, propelled by Love. For if all four capacities did not exist then individual existence would be a fraction only of what it now is. And further still, the incredible diversity of individuality is structured too such that Love can have its fullness by allowing interactions and the coming-together of myriad types of individuals expressing all the possibilities in Light. And so too it is with mega-organization and with the possibility of sustainable global civilization.

Each layer of subsequent complexity is as it were bought out to allow a more complete foundation for Love to express itself.

Chapter 8.2: All is Light

So we have seen that this universe is due to a play of light. Light slowed down and was able to exist in packets or quanta. In turn this allowed matter to be sustained, without dispersing right away, which has given concreteness to everything that we see around us.

But further, everything we see around us is nothing other than an emergence of the play of Light.

When we look at our hand it is in reality made up cells that comprise of nucleic acids, polysaccharides, lipids, and proteins. These are themselves forms of knowledge, power, harmony, and presence that are none other than a more dense functionality of the properties of Light.

But these molecular plans are themselves made from chains of molecules comprised primarily from oxygen, carbon, hydrogen, nitrogen, calcium, and phosphorus. The property of power – surfacing in hydrogen and calcium, and the property of knowledge – surfacing in oxygen, carbon, nitrogen, and phosphorous - exist in abundance in our bodies. But the trace elements manganese, iron, copper, zinc all derived from Light's property of presence – are there too and essential to the functioning of cells. And lanthanides that are a projection of Light's property of harmony are known to exist in the blood. So again we see that the whole functioning of the hand is dependent on Light within Light.

But looking in even deeper, every atom is itself made from the interaction of quarks, leptons, boson, and the Higgs-boson that are again emergences of Light's knowledge, power, harmony and presence properties.

So we see that our hand is itself a crystallization of several layers of Light.

And the Light did not diminish as it takes on these emergences. It is there in its fullness, just as the Ocean is present in its fullness in a wave or the spray that is flung from it. So our hand is all the fullness of creation and all the potentiality of creation present in the form of our hand.

But so then is everything else that we see around us, from the smallest to the largest. All is a tremendous play of Light and remembering That, the universe and all its creations take on a very different reality and meaning. For then we find that we are always with our Lover, our Illuminator, our Force, our Presence, our Harmonizer or Nurturer. There has never been any separation. The only separation has been because of the paucity or incompleteness of our vision.

Section 9: Complexification of Four-Foldness

Matter is positioned as developing due to a process of four-fold complexification. Paradoxically the development of matter does not stop with the emergence of substrates creating matter. Matter continues to develop even as non-material functionality expresses itself. This section will summarize this proposition.

Chapter 9.1 summarizes the application of the Light-Space-Time Emergence equation to all of emergence, and even to those states prior to emergence.

Chapter 9.2 summarizes four-fold complexification.

Chapter 9.1: Application of Light-Space-Time Emergence Equation - Pre-Big-Bang Existence to an Evolving-Universe

The Light-Space-Time Emergence equation (3.1.3) reproduced below for convenience essentially provides an algorithm by which Pre-Big-Bang existence proceeds to a reality of an Evolving-Universe.

$$Emergence_{light-space-time} = \left|\begin{array}{cc} \left[\begin{array}{c} C_{\infty}: [Pr, Po, K, H] \\ (\downarrow R_{C_K} = f(R_{C_{\infty}})) \\ C_K: [S_{Pr}, S_{Po}, S_K, S_H] \\ (\downarrow R_{C_N} = f(R_{C_K})) \\ C_{N}: f(S_{Pr}\ x\ S_{Po}\ x\ S_K\ x\ S_H) \\ (\downarrow R_{C_U} = f(R_{C_N})) \\ C_U: [P, V, M, C] \end{array}\right]_{Light} & \left[\begin{array}{c} M_3 \rightarrow System_X \\ (\uparrow F \rightarrow I) \\ M_2 \rightarrow S_{Systemx} \\ (\uparrow Sig \rightarrow F) \\ M_1 \rightarrow Sig_x \\ (\uparrow > P_x) \\ U \rightarrow x_U \end{array}\right]_{Space} \\ \left[\begin{array}{c} M_3 : -\infty \leq t \leq \infty \\ \downarrow \\ M_2 : 0 \geq t > \infty \\ \downarrow \\ M_1 : 0 > t > \infty \\ \downarrow \\ U \rightarrow \begin{array}{c} t \sim E_{Cell};\ TC: M_3 \rightarrow U \\ t \sim E_{Human};\ TC: U \rightarrow M_3 \end{array} \end{array}\right]_{Time} & TC \rightarrow x_T \end{array}\right|_{(x_U|x_T)}$$

Pre-Big-Bang Reality

Before the Big Bang it can be assumed, as per the discussion in Section 1, that only Light at ∞ existed. Hence (3.3.1) collapses to the highlighted portions only to create a modified version, Equation 9.1.1, Pre-Big-Bang Reality:

$$
Pre-Big-Bang
$$

$$
= \left[\begin{matrix} \left[\begin{matrix} C_{\infty}: [Pr, Po, K, H] \\ (\downarrow R_{C_K} = f(R_{C_\infty})) \\ C_K: [S_{Pr}, S_{Po}, S_K, S_H] \\ (\downarrow R_{C_N} = f(R_{C_K})) \\ C_{N:}\ f(S_{Pr}\ x\ S_{Po}\ x\ S_K\ x\ S_H) \\ (\downarrow R_{C_U} = f(R_{C_N})) \\ C_U:\ [P, V, M, C] \end{matrix}\right]_{Light} & \left[\begin{matrix} M_3 \rightarrow System_x \\ (\uparrow F \rightarrow I) \\ M_2 \rightarrow S_{System_x} \\ (\uparrow Sig \rightarrow F) \\ M_1 \rightarrow Sig_x \\ (\uparrow > P_x) \\ U \rightarrow x_U \end{matrix}\right]_{Space} \\ \left[\begin{matrix} M_3 : -\infty \leq t \leq \infty \\ \downarrow \\ M_2 : 0 \geq t > \infty \\ \downarrow \\ M_1 : 0 > t > \infty \\ \downarrow \\ U \rightarrow \begin{matrix} t \sim E_{Cell};\ TC: M_3 \rightarrow U \\ t \sim E_{Human};\ TC: U \rightarrow M_3 \end{matrix} \end{matrix}\right]_{Time} & TC \rightarrow x_T \end{matrix}\right]_{(x_U | x_T)}
$$

Eq 9.1.1: Pre-Big-Bang Reality

The Big-Bang

Assuming Light then projects itself from its state of ∞ to that of c instantaneously, this creates the Big Bang and yields Equation 9.1.2., Big Bang, where the highlighted portions only are operative. Note that essentially in this state the multiple quantization discussed in Section 3 becomes active.

$$
Big - Bang = \left[\begin{array}{cc}
\left[\begin{array}{c}
C_{\infty}: [Pr, Po, K, H] \\
(\downarrow R_{C_K} = f(R_{C_\infty})) \\
C_K: [S_{Pr}, S_{Po}, S_K, S_H] \\
(\downarrow R_{C_N} = f(R_{C_K})) \\
C_N: f(S_{Pr}\ x\ S_{Po}\ x\ S_K\ x\ S_H) \\
(\downarrow R_{C_U} = f(R_{C_N})) \\
C_U: [P, V, M, C]
\end{array} \right]_{Light}
&
\left[\begin{array}{c}
M_3 \rightarrow System_X \\
(\uparrow F \rightarrow I) \\
M_2 \rightarrow S_{System X} \\
(\uparrow Sig \rightarrow F) \\
M_1 \rightarrow Sig_x \\
(\uparrow > P_x) \\
U \rightarrow x_U
\end{array} \right]_{Space}
\\
\left[\begin{array}{c}
M_3 : -\infty \leq t \leq \infty \\
\downarrow \\
M_2 : 0 \geq t > \infty \\
\downarrow \\
M_1 : 0 > t > \infty \\
\downarrow \\
U \rightarrow \begin{array}{c} t \sim E_{Cell}; TC: M_3 \rightarrow U \\ t \sim E_{Human}; TC: U \rightarrow M_3 \end{array}
\end{array} \right]_{Time}
&
TC \rightarrow x_T
\end{array} \right]_{\langle x_U | x_T \rangle}
$$

Eq 9.1.2: Big-Bang

Space-Time-Energy-Gravity Emergence

As discussed in Chapter 3.5 Space, Time, Energy, and Gravity are emergent and are the result of Layer U being created with the 'settling' down of Light to a speed c. Hence Equation 9.1.3, Space-Time-Energy-Gravity Emergence is synonymous with (9.1.2):

$$
Space - Time - Energy - Gravity\ Emergence
$$

$$
= \left[\begin{array}{cc}
\left[\begin{array}{c}
C_{\infty}: [Pr, Po, K, H] \\
(\downarrow R_{C_K} = f(R_{C_\infty})) \\
C_K: [S_{Pr}, S_{Po}, S_K, S_H] \\
(\downarrow R_{C_N} = f(R_{C_K})) \\
C_N: f(S_{Pr} \times S_{Po} \times S_K \times S_H) \\
(\downarrow R_{C_U} = f(R_{C_N})) \\
C_U: [P, V, M, C]
\end{array} \right]_{Light}
&
\left[\begin{array}{c}
M_3 \rightarrow System_X \\
(\uparrow F \rightarrow I) \\
M_2 \rightarrow S_{System_X} \\
(\uparrow Sig \rightarrow F) \\
M_1 \rightarrow Sig_x \\
(\uparrow > P_x) \\
U \rightarrow x_U
\end{array} \right]_{Space}
\\
\left[\begin{array}{c}
M_3 : -\infty \leq t \leq \infty \\
\downarrow \\
M_2 : 0 \geq t > \infty \\
\downarrow \\
M_1 : 0 > t > \infty \\
\downarrow \\
U \rightarrow \begin{array}{c} t \sim E_{Cell};\ TC: M_3 \rightarrow U \\ t \sim E_{Human};\ TC: U \rightarrow M_3 \end{array}
\end{array} \right]_{Time}
&
TC \rightarrow x_T
\end{array} \right]_{\langle x_U | x_T \rangle}
$$

Eq 9.1.3: Space-Time-Energy-Gravity Emergence

It is with (9.1.3) that the emergence through matter, life, and mega-organization can proceed. For the Light-Matrix allows quantization to occur which means that the means for affecting the fabric of material existence is now in place.

As a reminder in the preceding adaptations of (3.1.3) the notion of evolution is captured by $\langle x_U | x_T \rangle$, which signifies x_T becoming the new starting point for x_U as the process of adaptability continues.

Hence, if we trace the potential in Light working from the basis of an initial electromagnetic spectrum to the current basis of a possible sustainable global civilization, some of the iterations the equation may have gone through can be envisioned as follows in Equation 9.1.4 – 10. Note that these are only snapshots and always there will be a vaster number of iterations that will be required.

Quantum Particle Emergence

In Equation 9.1.4, the starting point, x_U, is 'EM Spectrum', and the ending point, x_T, is 'Quantum Particles':

$$
\textit{Quantum Particle Emergence}
$$

$$
= \left\| \begin{array}{cc}
\left[\begin{array}{c}
C_{\infty}: [Pr, Po, K, H] \\
(\downarrow R_{C_K} = f(R_{C_\infty})) \\
C_K: [S_{Pr}, S_{Po}, S_K, S_H] \\
(\downarrow R_{C_N} = f(R_{C_K})) \\
C_{N:}\, f(S_{Pr}\, x\, S_{Po}\, x\, S_K\, x\, S_H) \\
(\downarrow R_{C_U} = f(R_{C_N})) \\
C_U: [P, V, M, C]
\end{array}\right]_{Light}
&
\left[\begin{array}{c}
M_3 \rightarrow System_x \\
(\uparrow F \rightarrow I) \\
M_2 \rightarrow S_{System_X} \\
(\uparrow Sig \rightarrow F) \\
M_1 \rightarrow Sig_x \\
(\uparrow > P_x) \\
U \rightarrow x_{EM\ Spectrum}
\end{array}\right]_{Space}
\\
\left[\begin{array}{c}
M_3 : -\infty \leq t \leq \infty \\
\downarrow \\
M_2 : 0 \geq t > \infty \\
\downarrow \\
M_1 : 0 > t > \infty \\
\downarrow \\
U \rightarrow \begin{array}{c} t \sim E_{Cell};\ TC: M_3 \rightarrow U \\ t \sim E_{Human};\ TC: U \rightarrow M_3 \end{array}
\end{array}\right]_{Time}
&
TC \rightarrow x_{Quantum\ particles}
\end{array} \right\|_{(x_U | x_T)}
$$

Eq 9.1.4: Quantum Particle Emergence

As discussed in detail in Section 2, Mathematical Foundations, the process of evolution proceeds through a constant breaking of 'patterns'. Section 3, and in particular Chapters 3.5 and 3.6 lays out the mathematical process of quantization effecting material reality. Sections 5, 6, and 7 further elaborated the process, which is only summarized in these series of equations in Chapter 9.

Atoms Emergence

In Equation 9.1.5, Atoms Emergence, the starting point, x_U, is 'Quantum Particles', and the ending point, x_T, is 'Atoms':

$$
\textit{Atoms Emergence} = \left|\begin{array}{cc}
\begin{bmatrix}
C_\infty: [Pr, Po, K, H] \\
(\downarrow R_{C_K} = f(R_{C_\infty})) \\
C_K: [S_{Pr}, S_{Po}, S_K, S_H] \\
(\downarrow R_{C_N} = f(R_{C_K})) \\
C_N: f(S_{Pr}\ x\ S_{Po}\ x\ S_K\ x\ S_H) \\
(\downarrow R_{C_U} = f(R_{C_N})) \\
C_U: [P, V, M, C]
\end{bmatrix}_{Light}
&
\begin{bmatrix}
M_3 \rightarrow System_X \\
(\uparrow F \rightarrow I) \\
M_2 \rightarrow S_{System X} \\
(\uparrow Sig \rightarrow F) \\
M_1 \rightarrow Sig_x \\
(\uparrow > P_x) \\
U \rightarrow x_{Quantum\ Particles}
\end{bmatrix}_{Space}
\\
\begin{bmatrix}
M_3 : -\infty \leq t \leq \infty \\
\downarrow \\
M_2 : 0 \geq t > \infty \\
\downarrow \\
M_1 : 0 > t > \infty \\
\downarrow \\
U \rightarrow \begin{matrix} t \sim E_{Cell};\ TC: M_3 \rightarrow U \\ t \sim E_{Human};\ TC: U \rightarrow M_3 \end{matrix}
\end{bmatrix}_{Time}
&
TC \rightarrow x_{Atoms}
\end{array}\right|_{\langle x_U | x_T \rangle}
$$

Eq 9.1.5: Atoms Emergence

Molecules Emergence

In Equation 9.1.6, Molecules Emergence, the starting point, x_U, is 'Atoms', and the ending point, x_T, is 'Molecules':

$$
\textit{Molecules Emergence} = \left|\begin{array}{cc}
\begin{bmatrix}
C_\infty: [Pr, Po, K, H] \\
(\downarrow R_{C_K} = f(R_{C_\infty})) \\
C_K: [S_{Pr}, S_{Po}, S_K, S_H] \\
(\downarrow R_{C_N} = f(R_{C_K})) \\
C_N: f(S_{Pr}\ x\ S_{Po}\ x\ S_K\ x\ S_H) \\
(\downarrow R_{C_U} = f(R_{C_N})) \\
C_U: [P, V, M, C]
\end{bmatrix}_{Light}
&
\begin{bmatrix}
M_3 \rightarrow System_X \\
(\uparrow F \rightarrow I) \\
M_2 \rightarrow S_{System X} \\
(\uparrow Sig \rightarrow F) \\
M_1 \rightarrow Sig_x \\
(\uparrow > P_x) \\
U \rightarrow x_{Atoms}
\end{bmatrix}_{Space}
\\
\begin{bmatrix}
M_3 : -\infty \leq t \leq \infty \\
\downarrow \\
M_2 : 0 \geq t > \infty \\
\downarrow \\
M_1 : 0 > t > \infty \\
\downarrow \\
U \rightarrow \begin{matrix} t \sim E_{Cell};\ TC: M_3 \rightarrow U \\ t \sim E_{Human};\ TC: U \rightarrow M_3 \end{matrix}
\end{bmatrix}_{Time}
&
TC \rightarrow x_{Molecules}
\end{array}\right|_{\langle x_U | x_T \rangle}
$$

Eq 9.1.6: Molecules Emergence

Cells Emergence

In Equation 9.1.7, Cells Emergence, the starting point, x_U, is 'Molecules', and the ending point, x_T, is 'Cells':

$$
\textit{Cells Emergence} = \left\| \begin{matrix} \left[\begin{matrix} C_\infty: [Pr, Po, K, H] \\ (\downarrow R_{C_K} = f(R_{C_\infty})) \\ C_K: [S_{Pr}, S_{Po}, S_K, S_H] \\ (\downarrow R_{C_N} = f(R_{C_K})) \\ C_{N:} f(S_{Pr}\ x\ S_{Po}\ x\ S_K\ x\ S_H) \\ (\downarrow R_{C_U} = f(R_{C_N})) \\ C_U: [P, V, M, C] \end{matrix} \right]_{Light} \left[\begin{matrix} M_3 \rightarrow System_X \\ (\uparrow F \rightarrow I) \\ M_2 \rightarrow S_{System X} \\ (\uparrow Sig \rightarrow F) \\ M_1 \rightarrow Sig_x \\ (\uparrow > P_x) \\ U \rightarrow x_{Molecules} \end{matrix} \right]_{Space} \\ \left[\begin{matrix} & M_3 : -\infty \leq t \leq \infty \\ & \downarrow \\ & M_2 : 0 \geq t > \infty \\ & \downarrow \\ & M_1 : 0 > t > \infty \\ & \downarrow \\ & t \sim E_{Cell}; TC: M_3 \rightarrow U \\ U \rightarrow & t \sim E_{Human}; TC: U \rightarrow M_3 \end{matrix} \right]_{Time} \quad TC \rightarrow x_{Cells} \end{matrix} \right\|_{(x_U | x_T)}
$$

Eq 9.1.7: Cells Emergence

Humans Emergence

In Equation 9.1.8, Humans Emergence, the starting point, x_U, is 'Cells', and the ending point, x_T, is 'Humans'. Clearly there will be a vast number of iterations before humans emerge:

$$
Humans\ Emergence = \left[\begin{matrix} \left[\begin{matrix} C_{\infty}: [Pr, Po, K, H] \\ (\downarrow R_{C_K} = f(R_{C_{\infty}})) \\ C_K: [S_{Pr}, S_{Po}, S_K, S_H] \\ (\downarrow R_{C_N} = f(R_{C_K})) \\ C_N: f(S_{Pr}\ x\ S_{Po}\ x\ S_K\ x\ S_H) \\ (\downarrow R_{C_U} = f(R_{C_N})) \\ C_U: [P, V, M, C] \end{matrix} \right]_{Light} & \left[\begin{matrix} M_3 \rightarrow System_X \\ (\uparrow F \rightarrow I) \\ M_2 \rightarrow S_{System X} \\ (\uparrow Sig \rightarrow F) \\ M_1 \rightarrow Sig_x \\ (\uparrow > P_x) \\ U \rightarrow x_{Cells} \end{matrix} \right]_{Space} \\ \left[\begin{matrix} M_3 : -\infty \leq t \leq \infty \\ \downarrow \\ M_2 : 0 \geq t > \infty \\ \downarrow \\ M_1 : 0 > t > \infty \\ \downarrow \\ U \rightarrow \begin{matrix} t \sim E_{Cell}; TC: M_3 \rightarrow U \\ t \sim E_{Human}; TC: U \rightarrow M_3 \end{matrix} \end{matrix} \right]_{Time} & TC \rightarrow x_{Humans} \end{matrix} \right]_{\langle x_U | x_T \rangle}
$$

Eq 9.1.8: Humans Emergence

If we follow the lines of emergence as per Sections 6 and 7 then (9.1.8) will be the starting point for potentially different trajectories of development. In one trajectory 'Humans' will evolve 'Basic Capacities of Self' and then 'Truer Individuality'. In another line of development 'Humans' will be the starting point for 'Mega-Organization' culminating in 'Sustainable Global Civilization'. The gist of these can be represented by Equations 9.1.9 and 9.1.10 respectively.

Truer Individuality Emergence

Hence, Equation 9.1.9, Truer Individuality Emergence, the starting point, x_U, is 'Humans', and the ending point, x_T, is 'Truer Individuality'. Clearly there will be a vast number of iterations before 'Truer Individuality' that will contain 'Basic Capacities of Self' as a milestone along the way:

$$
\mathit{Truer\ Individuality\ Emergence}
$$

$$
= \left| \begin{array}{ll}
\begin{bmatrix}
C_{\infty}: [Pr, Po, K, H] \\
(\downarrow R_{C_K} = f(R_{C_{\infty}})) \\
C_K: [S_{Pr}, S_{Po}, S_K, S_H] \\
(\downarrow R_{C_N} = f(R_{C_K})) \\
C_{N:} f(S_{Pr}\ x\ S_{Po}\ x\ S_K\ x\ S_H) \\
(\downarrow R_{C_U} = f(R_{C_N})) \\
C_U: [P, V, M, C]
\end{bmatrix}_{Light}
&
\begin{bmatrix}
M_3 \rightarrow System_X \\
(\uparrow F \rightarrow I) \\
M_2 \rightarrow S_{System\,X} \\
(\uparrow Sig \rightarrow F) \\
M_1 \rightarrow Sig_x \\
(\uparrow > P_x) \\
U \rightarrow x_{Humans}
\end{bmatrix}_{Space}
\\
\left[\begin{array}{c}
M_3 : -\infty \leq t \leq \infty \\
\downarrow \\
M_2 : 0 \geq t > \infty \\
\downarrow \\
M_1 : 0 > t > \infty \\
\downarrow \\
U \rightarrow \begin{array}{l} t \sim E_{Cell};\ TC: M_3 \rightarrow U \\ t \sim E_{Human};\ TC: U \rightarrow M_3 \end{array}
\end{array}\right]_{Time}
&
TC \rightarrow x_{Truer\ Individuality}
\end{array} \right|_{(x_U | x_T)}
$$

Eq 9.1.9: Truer Individuality Emergence

Sustainable Global Civilization Emergence

Equation 9.1.10, Sustainable Global Civilization Emergence, the starting point, x_U, is 'Humans', and the ending point, x_T, is 'Sustainable Global Civilization'. Clearly there will be a vast number of iterations before 'Sustainable Global Civilization' that will contain 'Mega-Organization as a milestone along the way:

$$
\textit{Sustainable Global Civilization Emergence} = \left| \begin{matrix} \begin{bmatrix} C_\infty: [Pr, Po, K, H] \\ (\downarrow R_{C_K} = f(R_{C_\infty})) \\ C_K: [S_{Pr}, S_{Po}, S_K, S_H] \\ (\downarrow R_{C_N} = f(R_{C_K})) \\ C_N: f(S_{Pr} \, x \, S_{Po} \, x \, S_K \, x \, S_H) \\ (\downarrow R_{C_U} = f(R_{C_N})) \\ C_U: [P, V, M, C] \end{bmatrix}_{Light} & \begin{bmatrix} M_3 \rightarrow System_X \\ (\uparrow F \rightarrow I) \\ M_2 \rightarrow S_{System_X} \\ (\uparrow Sig \rightarrow F) \\ M_1 \rightarrow Sig_x \\ (\uparrow > P_x) \\ U \rightarrow x_{Humans} \end{bmatrix}_{Space} \\ \begin{bmatrix} & M_3: -\infty \leq t \leq \infty \\ & \downarrow \\ & M_2: 0 \geq t > \infty \\ & \downarrow \\ & M_1: 0 > t > \infty \\ & \downarrow \\ U \rightarrow & t \sim E_{Cell}; TC: M_3 \rightarrow U \\ & t \sim E_{Human}; TC: U \rightarrow M_3 \end{bmatrix}_{Time} & TC \rightarrow x_{Sust.Global\ Civilization} \end{matrix} \right|_{(x_U | x_T)}
$$

Eq 9.1.10: Sustainable Global Civilization Emergence

Through a process such as summarized by the preceding continually adapting form of (3.1.3) from the emergence of Light as the EM Spectrum through a Sustainable Global Civilization there is a complexification of four-foldness. This is discussed in greater detail in the following chapter.

Self

Equations 9.1.1 through 9.1.10 that describe seemingly different realities starting from Pre-Big-Bang Existence, through the Big Bang, through emergence of space-time-energy-gravity, through the electromagnetic spectrum, quantum particles, atoms, cells, capacities of self, truer individuality, mega-organizations, and sustainable global civilization, are in reality only one and the same equation. This one equation that can describe everything and that is in everything is the Self. From it the Universe has emerged, the moorings that define the dynamics of the Universe have emerged, and then all of matter, all of life, and all of complex organization have emerged. This ONE equation, this SELF, has all the mathematics in it. It is one with all, and by making changes in itself it can change anything. This one equation is imbued with creative power, and such that it can create anything.

Chapter 9.2: Complexification of Four-Foldness

As per discussion in Section 1, light slowing down to c resulted in the birthing of matter in the phenomenon of the Big Bang.

In a common depiction of the history of the universe summarized by the Particle Data Group (Particle Data Group, 2015) at time, $t = 10^{-36}$ s, there was a high-intensity of light depicted by the cumulative energy of photons, $E = 10^{16}$ GeV. Existence of photons implies the existence of the EM Spectrum, and therefore tangibly, the existence of a fourfold wholeness projected from Light. As discussed previously in Section 4 this fourfold fullness was expressed by the wave-range of the EM Spectrum (Light's property of Knowledge), its implicit energy-gradient (Light's property of Power), the speed c with which is propagated (Light's property of Harmony), and the potential for mass (Light's property of Presence) also implicit in it.

But even before the emergence of the EM Spectrum, a more fundamental mooring of Space-Time-Energy-Gravity came into effect as discussed conceptually in Chapter 1.2 and mathematically in Chapter 3.5, and as summarized in the previous chapter. Given the proposition that Light has infinite variation on basic properties embedded in it, Space can be conceptualized as a seeding ground for the uniqueness implied in such infinite variation. Time can be thought of as a means by which these infinite seeds approach maturity. Energy can be thought of as some accumulation of the distinctness implied by seeds so that these may manifest in material form. Gravity can be thought of as the means by which relationship between such seeds is established.

In a 'Brief History of Time' Hawking (Hawking, 1988) states that the eventual goal of science is to provide a single theory that describes the whole universe. Scientists will usually separate this problem into two parts. First, focusing on laws that suggest how the universe changes with time. Second, focusing on the initial state of the universe. Hawking relates how some people feel that science should be concerned only with the first part. Since however, there appears to be very regular laws that govern how the universe changes with time, it must be equally reasonable to suppose that there are laws governing the initial state, he argues.

The existence of the four-foldness as an initial organizing principle is consistent with Hawking's argument. Only, in this mathematical model, it is the same set of laws that exists at both the initial starting point, and the subsequent development of the Universe, and as suggested in previous chapters. Chapter 9.1 illustrates how the Light-Space-Time Emergence equation (3.1.3) can frame dynamics from a pre-Big Bang existence to a continually evolving existence manifest thus far as a sustainable global civilization.

Space-Time-Energy-Gravity Four-Foldness

In looking at the nature of the fourfold order it is apparent that the fourfold wholeness is more tightly related to a smaller set of fundamental attributes at earlier stages of emergence.

Hence in thinking through the first apparent space-time-energy-gravity emergence, these may be summarized by the equations, Equations 9.2.1 through 9.2.4. In each case there is a dependence on 'seeds'.

Equation 9.2.1, System-Knowledge (Space) as Space-Time-Energy-Gravity Level, summarizes how knowledge is related to 'infinity of seeds' each representing as it were a different arrangement of knowledge:

$$System_{Knowledge} \propto [f(infinity\ of\ seeds)]$$

Eq 9.2.1: System-Knowledge (Space) at Space-Time-Energy-Gravity Level

Note that the infinity of seeds requires 'space' to be expressed.

Equation 9.2.2, System-Power (Time) at Space-Time-Energy-Gravity Level, summarizes how power is related to 'maturation of seeds', since anything that gets in the way of the maturation will generally be overcome:

$System_{Power} \propto [f(maturation\ of\ seeds)]$, where v is the frequency of the EM wave

Eq 9.2.2: System-Power (Time) at Space-Time-Energy-Gravity Level

Note that maturation of seeds is felt in 'time'.

Equation 9.2.3, System-Presence (Energy/Matter) at Space-Time-Energy-Gravity Level, summarizes how presence is related to 'materialization of seeds', since the accumulation of energy will express itself in material form:

$$System_{Presence} \propto [f(materialization\ of\ seeds)]$$

Eq 9.2.3: System-Presence (Energy/Matter) at Space-Time-Energy-Gravity Level

Equation 9.2.4, System-Harmony (Gravity) at Space-Time-Energy-Gravity Level, summarizes how harmony is related to 'relationship of seeds', since the accumulation of energy will express itself in material form:

$$System_{Harmony} \propto [f(relationship\ of\ seeds)]$$

Eq 9.2.4: System-Harmony (Gravity) at Space-Time-Energy-Gravity Level

Note that such relationship may be expressed as 'gravity'.

Electromagnetic (EM) Spectrum Four-Foldness

In the case of the EM spectrum the number of basic attributes to express four-foldness is two as will be observed.

Equation 9.2.5, System-Harmony at Electromagnetic Level, is related to the speed of light, c at U:

$$System_{Harmony} \propto c_U$$

Eq 9.2.5: System-Harmony at Electromagnetic Level

Equation 9.2.6, System-Power at Electromagnetic Level, is related to the frequency, ν, of the EM spectrum:

$$System_{Power} \propto hv$$

Eq 9.2.6: System-Power at Electromagnetic Level

Equation 9.2.7, System-Knowledge at Electromagnetic Level, is related to the wavelength, λ, of the EM spectrum:

$$System_{Knowledge} \propto [f(\lambda)]$$

Eq 9.2.7: System-Knowledge at Electromagnetic Level

And since, $E = mc^2$ or $m = \frac{E}{c^2}$, and substituting hv for E, yields Equation 9.2.8, System-Presence at Electromagnetic Level:

$$System_{Presence} \propto {hv}/{c^2}$$

Eq 9.2.8: System-Presence at Electromagnetic Level

But, $C \propto 1/h$

Hence, the general relationships of the four-fold wholeness may be understood by knowing just two variables, C, and either, λ or v.

Quantum Particle Four-Foldness

Now as per the model presented by the Particle Data Group at time, t = 10^{-10} s, the four-fold fullness begins to express itself in a series of particles. Quarks, leptons, bosons, which also may imply the presence of the Higgs-boson that gives quarks their mass are observed. But as discussed quarks are a precipitation of Light's property of Knowledge, leptons of Power, bosons of Harmony, and the Higgs-boson of Presence. Hence it is again observed that the four-fold fullness has expressed itself at a different order of complexity.

Einstein has suggested that particles are an excitation of the underlying field. The EM spectrum gives insight into a field that propagates at c at different frequencies. But there are other fields too. The Higgs Field has been suggested as being one of these (Jepsen, 2013). The Standard Model suggests that quarks, leptons, bosons, and the Higgs-boson are different fundamental particles, which implies that the fourfold wholeness has now a more complex basis of its operation, since as compared with the fourfold wholeness of the EM spectrum the number of independent bases appears to have increased as summarized in Figure 9.2.1, Complexification of Matter in Terms of Levels of Implicit Wholeness, to appear below.

Summarizing, as in Equations 9.2.9 through 9.2.12:

$$System_{Harmony} \propto f(bosons)$$

Eq 9.2.9: System-Harmony at Particle Level

$$System_{Power} \propto f(leptons)$$

Eq 9.2.10: System-Power at Particle Level

$$System_{Knowledge} \propto f(quarks)$$

Eq 9.2.11: System-Knowledge at Particle Level

$$System_{Presence} \propto f(Higgs\_boson)$$

Eq 9.2.12: System-Presence at Particle Level

Periodic Table/Atoms Four-Foldness

As discussed in Chapter 5.3 the Periodic Table itself is also an expression of the four-fold wholeness. Hence, d-group elements are an instance of Light's property of Presence, p-group elements are an instance of Light's property of Knowledge, s-group elements are an instance of Light's property of Power, and f-group elements are an instance of Light's property of Harmony or Nurturing. As per the Particle Data Group model (Particle Data Group, 2015) this emergence likely began at $t = 3 \times 10^5$ years when the atom is suggested to have emerged, and continued at least to $t = 10^9$ years with the emergence of stars which it is already known are the furnaces in which heavier atoms were created.

Leveraging the fundamentally different groups yields Equations 9.2.13 through 9.2.16:

$$System_{Harmony} \propto f(f\_group)$$

Eq 9.2.13: System-Harmony at Atomic Level

$$System_{Power} \propto f(s\_group)$$

Eq 9.2.14: System-Power at Atomic Level

$$System_{Knowledge} \propto f(p\_group)$$

Eq 9.2.15: System-Knowledge at Atomic Level

$$System_{Presence} \propto f(d\_group)$$

Eq 9.2.16: System Presence at Atomic Level

Atoms therefore have at least four four-fold wholenesses implicit in them.

Living Cell Four-Foldness

At time, t = 13.8 x 10^9 years, there is a fifth clear expression of the same four-fold order as the bases of an even more complex organization, that of cellular life and all that is founded on it. This can be summarized as in Equations 9.2.17 through 9.2.20:

$$System_{Harmony} \propto f(lipids)$$

Eq 9.2.17: System-Harmony at Cellular Level

$$System_{Power} \propto f(polysaccharides)$$

Eq 9.2.18: System-Power at Cellular Level

$$System_{Knowledge} \propto f(nucleic\ acids)$$

Eq 9.2.19: System-Knowledge at Cellular Level

$$System_{Presence} \propto f(proteins)$$

Eq 9.2.20: System-Presence at Cellular Level

Cellular life therefore seems to have at least five four-fold-wholenesses implicit in it to therefore further complexify the architecture of matter.

Fundamental Capacities of Self Four-Foldness

In the mathematical model elaborated in this book, due to Space-Time-Energy-Gravity (S-T-E-G) quantization the reality of matter is always changing. Such quantization is the result of quantum-certainty as discussed in Chapter 3.5, which itself is the result of a cohesive want or will. Since all is Light, as also discussed, Light therefore can change Light by Light. Deductively therefore, a changing matter is inevitable.

Further, as S-T-E-G quantization proposes matter will be changed by seemingly non-physical phenomena. Hence, as new function-elements from the underlying four architectural sets express themselves or are created, matter itself will go through a change. Fundamental capacities of self, hence, will alter the material fabric or the reality of matter.

Up to the level of living cells there are at least five four-fold-wholenesses implicit in it. As will be proposed here and in subsequent sub-sections in this chapter the number of these implicit four-fold-wholenesses will continue to increase.

Equations 9.2.21 through 9.2.24 summarizes a sixth clear expression of four-foldness that will change the structure of matter.

$$System_{Harmony} \propto f(Emotions, Feelings)$$

Eq 9.2.21: System-Harmony at Level of Capacities of Self

$$System_{Power} \propto f(Urges, Wills, Desires)$$

Eq 9.2.22: System-Power at Level of Capacities of Self

$$System_{Knowledge} \propto f(Thoughts)$$

Eq 9.2.23: System-Knowledge at Level of Capacities of Self

$$System_{Presence} \propto f(Sensations)$$

Eq 9.2.24: System-Presence at Level of Capacities of Self

Truer Individuality Four-Foldness

Similarly as truer individuality expresses itself S-T-E-G quantization initiated by quantum-certainty will further complexify the bases of matter.

Equations 9.2.25 through 9.2.28 therefore summarize a seventh clear expression of four-foldness that will change the structure of matter.

$$System_{Harmony} \propto f(Harmony - type - individuals)$$

Eq 9.2.25: System-Harmony at Level of Truer Individuality

$$System_{Power} \propto f(Power \quad type \quad individuals)$$

Eq 9.2.26: System-Power at Level of Truer Individuality

$$System_{Knowledge} \propto f(Knowledge - type - individuals)$$

Eq 9.2.27: System-Knowledge at Level of Truer Individuality

$$System_{Presence} \propto f(Service - type - individuals)$$

Eq 9.2.28: System-Presence at Level of Truer Individuality

Sustainable Global Civilization Four-Foldness

Similarly as more complex organizational levels, such as a sustainable global civilization expresses itself, S-T-E-G quantization initiated by quantum-certainty will further complexify the bases of matter. Recall this now brings into matter the capacity represented by creating a sustainable global civilization into its fold as it were to be used consciously or sub-consciously by matter-based entities going forward.

Equations 9.2.29 through 9.2.32 therefore summarize an eighth clear expression of four-foldness that will change the structure of matter. Note that these expressions are indicative only and there are other missed levels that would need to be expressed for completeness.

$$System_{Harmony} \propto f(Harmony - type - region)$$

Eq 9.2.29: System-Harmony at Level of Nations or Blocs

$$System_{Power} \propto f(Power - type - region)$$

Eq 9.2.30: System-Power at Level of Nations or Blocs

$$System_{Knowledge} \propto f(Knowledge - type - region)$$

Eq 9.2.31: System-Knowledge at Level of Nations or Blocs

$$System_{Presence} \propto f(Service - type - region)$$

Eq 9.2.32: System-Presence at Level of Nations or Blocs

Summary of Complexification

This scheme of complexification is summarized by Figure 9.2.1:

Level	**Independent "Variables"**	**Minimum Levels of Implicit Wholeness**	**Minimum Number of Levels**
Space-Time-Gravity-Energy (S-T-G-E)	Seeds	• S-T-G-E	1
EM Spectrum	C, ν or λ	• S-T-G-E • EM	2
Particle Level	Quarks, Leptons,	• S-T-G-E	3

	Bosons, Higgs-Boson	• EM • Particle	
Atomic Level	s-group, p-group, d-group, f-group	• S-T-G-E • EM • Particle • Atom	4
Cellular Level	Proteins, Nucleic Acids, Lipids, Polysaccharides	• S-T-G-E • EM • Particle • Atom • Cell	5
Level of Capacities of Self	Sensations, Wills, Emotions, Thoughts	• S-T-G-E • EM • Particle • Atom • Cell • Capacities of Self	6
Truer Individuality Level	Service-type-individual, Knowledge-type-individual, Harmony-type-individual, Power-type-individual	• S-T-G-E • EM • Particle • Atom • Cell • Capacities of Self • Truer Individuality	7
Sustainable Global Civilization Level	Service-type-region, Knowledge-type-region, Harmony-type-region, Power-type-region	• S-T-G-E • EM • Particle • Atom • Cell • Capacities of Self • Truer Individuality • Sustainable Global Civilization	8

Figure 9.2.1 Complexification of Matter in Terms of Levels of Implicit Wholeness

In reference to the four sets at M_2 (as discussed in Chapter 2.3 on Architectural Forces) it may also be that as the basis of matter complexifies as it journeys through the EM spectrum and other potential fields, the elementary particles, the atom, cellular life, complex individual and further organizational development, as summarized in Figure 9.2.1, the sets which have been positioned to each contain infinite elements, concretely manifest more of their function-elements.

This suggestion has been the basis of the equations derived in previous chapters. The complexity of matter may therefore be related to the number of manifested function-elements of the set of four sets. This idea is consistent with the notion of the "adjacent possible" suggested by Kaufmann (Kaufmann, 2003) in which innovation is positioned as a recombination of existing parts to create new value – or of existing sets to combine parts of themselves to create new elements based on new circumstance. If MS signifies manifested-set, so that the cardinality or number of elements in the combined set is the union of the four manifested-sets, this may be summarized by the following equation, Equation 9.2.33:

$$Matter_{Complexity} \propto \left| MS_{System_{Pr}} \cup MS_{System_P} \cup MS_{System_K} \cup MS_{System_N} \right|$$

Eq 9.2.33: Complexity of Matter in Terms of Manifested Set

Section 10: The Making of History

Leveraging the previously developed quantum mathematical model this chapter will elaborate how history is made.

Chapter 10.1 focuses on the making of Natural History through considering the case of the adaptation of an insect.

Chapter 10.2 will elaborate how Sustainable Human History is approached through considering the case of Nations operating from a space of their deeper essence as a step towards the creation of a sustainable global civilization.

Chapter 10.1: The Making of Natural History

Leveraging the previously developed quantum mathematical model this chapter will elaborate how Natural History is made through considering the case of the adaptation of an insect.

The quantum is a window into layers of reality behind the surface layer U. This insight was captured by Schrodinger's Equation (Chapter 3.3) and by Heisenberg's Uncertainty Principle (Chapter 3.4), the former suggesting that matter, or what is going to appear materially, is a function of some incredible number of superposed states containing infinite possibility, and the latter suggesting that quantum fluctuation due to an always buzzing pregnant-infinity, lies behind everything even when seemingly still.

The Light-Space-Time Emergence equation (3.1.3) derived in Chapter 3.1, and reproduced here for convenience, may provide some insight into the dynamics of the pregnant-infinity, and of quantum-certainty in the adaptabilities of the cell. Recall that the basis of cell adaptability was discussed in Chapter 6.1. Reproducing the equation:

$$Emergence_{light-space-time}$$

$$= \left| \begin{array}{cc} \left[\begin{array}{c} C_\infty: [Pr, Po, K, H] \\ (\downarrow R_{C_K} = f(R_{C_\infty})) \\ C_K: [S_{Pr}, S_{Po}, S_K, S_H] \\ (\downarrow R_{C_N} = f(R_{C_K})) \\ C_N: f(S_{Pr}\ x\ S_{Po}\ x\ S_K\ x\ S_H) \\ (\downarrow R_{C_U} = f(R_{C_N})) \\ C_U: [P, V, M, C] \end{array} \right]_{Light} & \left[\begin{array}{c} M_3 \rightarrow System_X \\ (\uparrow F \rightarrow I) \\ M_2 \rightarrow S_{System X} \\ (\uparrow Sig \rightarrow F) \\ M_1 \rightarrow Sig_x \\ (\uparrow > P_x) \\ U \rightarrow x_U \end{array} \right]_{Space} \\ \left[\begin{array}{cc} & M_3 : -\infty \leq t \leq \infty \\ & \downarrow \\ & M_2 : 0 \geq t > \infty \\ & \downarrow \\ & M_1 : 0 > t > \infty \\ & \downarrow \\ U \rightarrow & \begin{array}{c} t \sim E_{Cell};\ TC: M_3 \rightarrow U \\ t \sim E_{Human};\ TC: U \rightarrow M_3 \end{array} \end{array} \right]_{Time} & TC \rightarrow x_T \end{array} \right|_{(x_U | x_T)}$$

The Light-Matrix, the top-left matrix in (3.1.3), suggests the possibilities in the pregnant-infinity. These possibilities, recall, have been set up by realities so created by light traveling at different speeds.

While prevalent dynamics at the surface layer, U, are always influenced by dynamics from the deeper layers, there is a state that can be created that will allow a more focused and intentional activation of the meta-functions resident in the deeper layers. Such activation will potentially trigger a series of processes culminating in a state of quantum-certainty, introduced in Chapter 3.5 when taking a deeper look at S-T-E-G quantization, and as will be explored through the rest of this chapter.

The Space-Matrix, the top-right matrix in (3.1.3), suggests the dynamics involved in creating such an activation-state. Essentially this involves the overcoming of patterns at U.

Insect Adaptation Case

Consider the case of a hypothetical insect, for example, that is always prey to other predators. At some point a visceral urge to overcome some of these predators may arise in the insect. This visceral urge is a breaking of habitual patterns common to that species of insect. If it is deep enough and pervasive enough, it can be thought of as an activation-state and will allow the opening of a quantum-window so that at every micro level there is now a connection with the quantum worlds, Q, behind.

It is proposed that for effective adaptability to come about it is only such a pervasive state that will open a quantum-window to effectively stimulate the interaction between layers that will ultimately allow adaptability to occur. This activation-state can be further specified by leveraging (2.7.4) – The Generalized Equation for Organizational Direction, derived in Chapter 2.7, on Qualified Determinism.

Reproducing Equation 2.7.4:

$$Org\_Dir = DI\left(\begin{bmatrix} M_3 \rightarrow System_X \\ (\uparrow F \rightarrow I) \\ M_2 \rightarrow S_{System_X} \\ (\uparrow Sig \rightarrow F) \\ M_1 \rightarrow Sig_x \\ (\uparrow > P_P) \\ U \rightarrow x_U \end{bmatrix}_{x=p,v,m,i}\right) \rightarrow$$

$$x\_matrix_{strongest} \;@\; level_{strongest}$$

The activation-state can be thought of as being invoked when $level_{strongest}$ is at least M_1. Hence, as in Equation 10.1.1:

$$Activation - State_{condition} \geq M_1$$

Eq 10.1.1: Condition for Activation-State

Note that the Time-Matrix, in the lower-left side of (3.1.3), specifies that approximately, adaptability may tend to be more automatic up to the emergence of cellular organisms. This is specified by the direction $M_3 \rightarrow U$, which indicates that it is the meta-level that organizes activity at U. At the human level the direction is flipped as specified by $U \rightarrow M_3$, and suggests that will, intention, feeling and the like are more important in stimulating the process of adaptability.

The visceral urge, then, may stimulate interaction with a collective-intelligence or 'specific-species-intelligence meta-function' that allows the generation of a new and specific 'predator-overcoming meta-function' so that the hypothetical insect in question goes through an adaptation to survive at least some kinds of predator attacks. This may take the form of this species of insect creating a hard-shell around it. Mathematically this new meta-function may take the following form as suggested by Equation 10.1.2:

$$Sig_{hard-protective-shell} = Xa + \overline{Yb_{0-n}} \quad where \left[\begin{array}{c} X \in [S_{System_P}] \\ Y \in [S_{System_{Pr}}, S_{System_P}, S_{System_K}, S_{System_N}] \\ a, b \ are \ integers; a > b \end{array} \right]$$

Eq 10.1.2: Hard Protective Shell Meta-Function

Specifically, S_{System_P} relates to the set of Polysaccharides and suggests that the chains of sugar molecules will adapt to become a shell to protect the insect. The primary element X, is therefore an element of the set or Polysaccharides. $S_{System_{Pr}}$ refers to the set of Proteins. S_{System_K} refers to the set of Nucleic Acids. S_{System_N} refers to the set of Lipids. Y as the secondary element will invoke the action of some proteins, some existing polysaccharides, some nucleic acids, and some lipids in bringing about the adaptation as specified by $Sig_{hard-protective-shell}$. Perhaps the nucleic acids will code or coordinate how the proteins will work with the existing polysaccharides and lipids to create a new arrangement of polysaccharides that becomes the protective layer for the insect. This new arrangement of polysaccharides then becomes the purpose of a particular new type of specialized cell that exists to protect the insect against certain types of predators.

So while the breaking of patterns as specified by the Space-Matrix was a first step in the process, the subsequent creation of a new meta-function can be thought of as a second step. This new meta-function can be thought of as happening due to some combined functioning of the Light and Space Matrices. Specifically in the Space-Matrix, the breaking of patterns, $(\uparrow > P_x)$, allows the dynamics of M_1 to become active: $M_1 \rightarrow Sig_x$. This allows the ever-present layers of light, and specifically $C_N: f(S_{Pr} \, x \, S_{Po} \, x \, S_K \, x \, S_H)$ in the Light-Matrix to become active so that new meta-function is generated.

The third step is suggested by $(\downarrow R_{C_U} = f(R_{C_N}))$ that allows specific quantization to occur. It is proposed that the hard-protective-shell adaptability becomes real through a series of tightly coordinated space, time, energy, and gravity quantization as specified by (3.5.2-5) derived in Chapter 3.5. These equations are reproduced here for convenience and suggest the mechanics for the adaptability along this specific line of development in the larger scheme of Natural History.

Equation 3.5.2 models space-quantization:

$$Space_{quantization} = h_{UK}(Xa + Y\bar{b}_{0-n})$$

$$where: \left[\begin{array}{c} X \in [S_{System_K}] \\ Y \in [S_{System_{Pr}}, S_{System_P}, S_{System_K}, S_{System_N}] \\ a, b \text{ are integers}; a > b \end{array} \right]$$

Recall that space-quantization further structures space so that seeds of knowledge or knowledge-potential to do with the new meta-function are now resident in the material ecosystem associated with the meta-function. In such a manner the species that set this meta-function into action will more and more tap into the possibilities to bring about the new protective mechanism it seeks.

Equation 3.5.3 models time-quantization:

$$Time_{quantization} = h_{UP}(Xa + Y\bar{b}_{0-n})$$

$$where: \left[\begin{array}{c} X \in [S_{System_P}] \\ Y \in [S_{System_{Pr}}, S_{System_P}, S_{System_K}, S_{System_N}] \\ a, b \text{ are integers}; a > b \end{array} \right]$$

Recall that time-quantization alters the structure of time so that phases of maturity associated with seeds in space are now encoded in time. Slow and faster periods of change are in this way connected to the species seeking to protect itself against predators.

Equation 3.5.4 models energy-quantization:

$$Energy_{quantization} = h_{UPr}(Xa + Y\bar{b}_{0-n})$$

$$where: \begin{bmatrix} X \in [S_{System_{Pr}}] \\ Y \in [S_{System_{Pr}}, S_{System_{P}}, S_{System_{K}}, S_{System_{N}}] \\ a, b \text{ are integers}; a > b \end{bmatrix}$$

Recall that energy-quantization allows matter to be formed, and in this case, will have to do with the incremental material changes that the species will go through in forming a hard protective shell.

Equation 3.5.5 models gravity-quantization:

$$Gravity_{quantization} = h_{UN}(Xa + Y\bar{b}_{0-n})$$

$$where: \begin{bmatrix} X \in [S_{System_{N}}] \\ Y \in [S_{System_{Pr}}, S_{System_{P}}, S_{System_{K}}, S_{System_{N}}] \\ a, b \text{ are integers}; a > b \end{bmatrix}$$

Recall that gravity-quantization has to do with inter-relation between the species and surrounding objects and will change the very nature of gravity to allow a subtle new balance in the species interaction with its surrounding so that the deep urge of the species can more easily be fulfilled.

Note that the previously derived equation (3.6.5) - Quantization Effect of Organization on Material Fabric – may also be leveraged to understand a possible quantization effect of organization on the material fabric. Reproducing (3.6.5):

Impact on Materal Fabric

$$= \begin{bmatrix} |[L]|[S]|[T]TC \rightarrow x_T|_{(x_U|x_T)} \\ \times \\ \left(|mod(Z_Q)|_{Y>U}\right) \\ \ni \\ Z_Q, Z \in \mathbb{U}\ (Space, Time, Energy, Gravity) \end{bmatrix}$$

Line 1 in the matrix is simply (3.6.1) the Simplified Light-Space-Time Emergence equation. To understand if quantization is going to occur, the organization resulting from the application of Line 1 is, where 'is' means 'multiplied' or ($\times$), by a 'modulation', specified by (mod), of a fourfold space-time-energy-gravity quantization (Z_Q) kind, so long as Y>U, that is, the operative bases are relatively transformed and hence have an active influence greater than U. The fourfold quantization is specified by $(\ni Z_Q, Z \in U\ (Space, Time, Energy, Gravity))$ where $(\ni)$ the quantization being applied (Z) is each (U) of the elements of the set (Space, Time, Energy, Gravity).

Such quantization of space, time, energy, and gravity, describes the state of quantum-certainty. Quantum-certainty is the result of an intimate interaction with Q, the worlds accessed through quanta. The process of quantization and the state of quantum-certainty initiated through an activation-state, and the consequent enhancement or creation of a new meta-function, give a more intimate sense of the tightly coherent system we live in. Urge or desire, if deep enough, can set into motion an inter-relation between easily distinct layers of existence, allowing a deeper 'system-being' to intervene in regular system functioning.

It is proposed that all of Natural History adapted and evolved in such a manner.

Leveraging the previously developed quantum mathematical model this chapter will elaborate how Sustainable Human History is approached through considering the case of Nations operating from a space of their deeper essence as a step towards the creation of a sustainable global civilization.

Just as matter can be further complexified to accommodate enhanced functionality as described in the example of an insect in the previous chapter, thereby also impacting natural history, so too will matter continue to change with the application of human will and want, thereby changing human history. This chapter describes the process of quantization by which this might occur.

The Light-Space-Time Emergence equation (3.1.3) derived in Chapter 2.8, and reproduced here for convenience, will provide some insight into the dynamics of the pregnant-infinity, and of quantum-certainty in the making of history. Reproducing the equation:

$$Emergence_{light-space-time} = \left\| \begin{array}{cc} \begin{bmatrix} C_\infty: [Pr, Po, K, H] \\ (\downarrow R_{C_K} = f(R_{C_\infty})) \\ C_K: [S_{Pr}, S_{Po}, S_K, S_H] \\ (\downarrow R_{C_N} = f(R_{C_K})) \\ C_N: f(S_{Pr}\ x\ S_{Po}\ x\ S_K\ x\ S_H) \\ (\downarrow R_{C_U} = f(R_{C_N})) \\ C_U: [P, V, M, C] \end{bmatrix}_{Light} & \begin{bmatrix} M_3 \to System_x \\ (\uparrow F \to I) \\ M_2 \to S_{System_x} \\ (\uparrow Sig \to F) \\ M_1 \to Sig_x \\ (\uparrow > P_x) \\ U \to x_U \end{bmatrix}_{Space} \\ \begin{bmatrix} M_3 : -\infty \le t \le \infty \\ \downarrow \\ M_2 : 0 > t > \infty \\ \downarrow \\ M_1 : 0 > t > \infty \\ \downarrow \\ t \sim E_{Cell};\ TC: M_3 \to U \\ U \to \quad t \sim E_{Human};\ TC: U \to M_3 \end{bmatrix}_{Time} & TC \to x_T \end{array} \right\|_{\langle x_U | x_T \rangle}$$

The Light-Matrix, the top-left matrix in (3.1.3), suggests the possibilities in the pregnant-infinity. These possibilities, recall, have been set up by realities so created by light traveling at different speeds. While prevalent dynamics at the surface layer, U, are always influenced by dynamics from the deeper layers, there is an activation-state that can be created that will allow a more focused and intentional activation or creation of the meta-functions in the deeper layers.

The Space-Matrix, the top-right matrix in (3.1.3), suggests the dynamics involved in creating such an activation-state. Essentially this involves the overcoming of patterns at U. Note that by definition such an activation-state necessitates moving in the direction where the speed of light increases. Note that speed of light increasing implies that the sense of separation is diminishing, and entities can have a more complete access to the fullness that they are. This can only happen when patterns get elevated. If patterns are in the U realm, being therefore untransformed, that will by definition not allow a quantum-window to open. Therefore the impact to history is termed as 'sustainable human history'.

As per discussion in Chapter 6.2, this activation-state can be further specified by leveraging (2.7.4) – The Generalized Equation for Organizational Direction, derived in Chapter 2.7, on Qualified Determinism.

Reproducing Equation 2.7.4:

$$Org\_Dir = DI\left(\begin{bmatrix} M_3 \rightarrow System_X \\ (\uparrow F \rightarrow I) \\ M_2 \rightarrow S_{System_X} \\ (\uparrow Sig \rightarrow F) \\ M_1 \rightarrow Sig_x \\ (\uparrow > P_{P)} \\ U \rightarrow x_U \end{bmatrix}_{x=p,v,m,i}\right) \rightarrow$$

$$x\_matrix_{strongest} @ level_{strongest}$$

Activation-state can be thought of as being invoked when $level_{strongest}$ is at least M_1. Hence, as suggested by the previously derived (10.1.1), reproduced here for convenience:

$$Activation - state_{condition} \geq M_1$$

Deeper Nation Essence Case

As per the discussion in the previous chapter, assume that there is a deep wanting felt by a threshold number of people. Perhaps this wanting, a step towards a global sustainable civilization, is for each nation to operate more from their true essence, or I-state, as opposed to fundamentalist form. In other words a meta-function, $Sig_{Nation-essence}$, must be created that will then organize its own reality, and as indicated by Equation 10.2.1.

$$Sig_{Nation-essence} = Xa + Y\bar{b}_{0-n}$$

$$where \begin{bmatrix} X \in [S_{System_K}] \\ Y \in [S_{System_{Pr}}, S_{System_P}, S_{System_K}, S_{System_N}] \\ a, b \text{ are integers}; a > b \end{bmatrix}$$

Eq 10.2.1: Nation-Essence Meta-Function

While the key aspect, or primary X-element in the nation-essence meta-function may be an element from the set of Knowledge, and specifically 'what the nation-essence means', there will be many parts, represented by the Y-elements, such as the structure, processes, institutions, culture, that will also need to be created. An important denominator in breaking patterns involves the creation of 'cohesion' at the individual and collective levels as in Equation 10.2.2, Nation-Essence-Cohesion with the primary element belonging to the set of Harmony or Nurturing:

$$Sig_{Nation-essence-cohesion} = Xa + Y\bar{b}_{0-n}$$

$$where \begin{bmatrix} X \in [S_{System_N}] \\ Y \in [S_{System_{Pr}}, S_{System_P}, S_{System_K}, S_{System_N}] \\ a, b \text{ are integers}; a > b \end{bmatrix}$$

Eq 10.2.2: Nation-Essence-Cohesion Meta-Function

Hence, S_{System_K} relates to the set of Thoughts required to contextualize and frame nation-essence. $S_{System_{Pr}}$ refers to the set of

Sensations required to assess and interpret the constantly arising signs of the new development – the sensory cues as it were that will allow any individual or collectivity to sense that they were on the right path. S_{System_P} refers to the set of Urges and Wills constantly required to ensure that the goal of cohesion was attained. S_{System_N} refers to the set of Feelings and Emotions that will need to be generated to attain cohesion.

As such cohesion becomes a reality, older patterns will break down and there will be easier and longer activation-state periods, potentially allowing 'miracle' to become the law.

As suggested in the previous chapter, while the breaking of patterns as specified by the Space-Matrix was a first step in the process, the subsequent creation of a new meta-function can be thought of as a second step. This new meta-function can be thought of as happening due to some combined functioning of the Light and Space Matrices. Specifically in the Space-Matrix, the breaking of patterns, $(\uparrow > P_x)$, allows the dynamics of M_1 to become active: $M_1 \rightarrow Sig_x$. This allows the ever-present layers of light, and specifically $C_{N:} f(S_{Pr} \, x \, S_{Po} \, x \, S_K \, x \, S_H)$ in the Light-Matrix to become active so that new meta-function is generated.

The third step is suggested by $(\downarrow R_{C_U} = f(R_{C_N}))$ that allows specific quantization to occur. Quantization is of fundamental importance because it is what allows the fabric of experienced reality to change. The new thought to do with nation-essence is quantized as seeds of knowledge in space, fundamentally changing the structure of space. Hence, reproducing Equation 3.5.2, space-quantization is modeled as:

$$Space_{quantization} = h_{UK}(Xa + Y\bar{b}_{0-n})$$

$$where: \begin{bmatrix} X \in [S_{System_K}] \\ Y \in [S_{System_{Pr}}, S_{System_P}, S_{System_K}, S_{System_N}] \\ a, b \; are \; integers; a > b \end{bmatrix}$$

The phases of maturity that the seeds of knowledge will go through give time a new reality or structure. In other words time too is re-oriented to promote the outcome of the new meta-function in effect. This quantization of time, essentially giving time in this new fabric a different meaning is modeled by Equation 3.5.3, reproduced here for convenience:

$$Time_{quantization} = h_{UP}(Xa + Y\bar{b}_{0-n})$$

$$where: \begin{bmatrix} X \in [S_{System_P}] \\ Y \in [S_{System_{Pr}}, S_{System_P}, S_{System_K}, S_{System_N}] \\ a, b\ are\ integers; a > b \end{bmatrix}$$

The new meta-function changes the very nature of matter, which is quantized to allow the material fabric of existence to promote the new meta-function in unimaginable ways. This quantization of energy, resulting in a subtly different matter is modeled by Equation 3.5.4, reproduced here for convenience:

$$Energy_{quantization} = h_{UPr}(Xa + Y\bar{b}_{0-n})$$

$$where: \begin{bmatrix} X \in [S_{System_{Pr}}] \\ Y \in [S_{System_{Pr}}, S_{System_P}, S_{System_K}, S_{System_N}] \\ a, b\ are\ integers; a > b \end{bmatrix}$$

As a result the very force of gravity is altered locally as well, so that its quantization promotes subtle interaction in the related ecosystem that will be predisposed to making the meta-function a reality. This quantization of gravity is modeled by Equation 3.5.5, reproduced here for convenience:

$$Gravity_{quantization} = h_{UN}(Xa + Y\bar{b}_{0-n})$$

$$where: \begin{bmatrix} X \in [S_{System_N}] \\ Y \in [S_{System_{Pr}}, S_{System_P}, S_{System_K}, S_{System_N}] \\ a, b\ are\ integers; a > b \end{bmatrix}$$

Note that the previously derived equation (3.6.5) - Quantization Effect of Organization on Material Fabric – may also be leveraged to understand a possible quantization effect of organization on the material fabric. Reproducing (3.6.5):

Impact on Materal Fabric

$$= \begin{bmatrix} |[L][S][T]TC \rightarrow x_T|_{(x_U|x_T)} \\ \times \\ \left(|mod(Z_Q)|_{Y>U}\right) \\ \ni \\ Z_Q, Z \in \mathbb{U}\ (Space, Time, Energy, Gravity) \end{bmatrix}$$

Line 1 in the matrix is simply (3.6.1) the Simplified Light-Space-Time Emergence equation. To understand if quantization is going to occur, the organization resulting from the application of Line 1 is, where 'is' means 'multiplied' or ($\times$), by a 'modulation', specified by (mod), of a fourfold space-time-energy-gravity quantization (Z_Q) kind, so long as Y>U, that is, the operative bases are relatively transformed and hence have an active influence greater than U. The fourfold quantization is specified by ($\ni Z_Q, Z \in U\ (Space, Time, Energy, Gravity)$) where ($\ni$) the quantization being applied (Z) is each (U) of the elements of the set (Space, Time, Energy, Gravity).

It is the combined quantization that can be thought of as quantum-certainty. Such a process of quantum-certainty has the effect of changing the vary fabric or make-up of matter and thereby creating a foundation for sustainable human history.

It is proposed that all of Sustainable Human History will adapt and evolve in such a manner.

Section 11: Transitioning to Super-Matter and the Expansion of the Universe

Chapter 11.1 discusses the creation of super-matter from a macro perspective. Super-matter is suggested to be that type of matter created through the intervention of conscious will or cohesive want. While complexification of matter as summarized in Section 10 is often an "automatic" process of Nature, super-matter is a foundation based on will or cohesive want and sets the stage for a potentially unending willed development in which functional-richness existing in Light at Infinity can manifest in this material universe. From a macro-level, meta-level bases are what will ensure the development of super-matter.

Chapter 11.2 draws a possible relationship between an expanding universe and the development of functionally-rich matter. As suggested creation of such matter requires space, time, energy, and gravity quantization which further implies that space as the prerequisite seeding ground of function must continue to expand.

Chapter 11.1: Transitioning to Super-Matter

Super-Matter is positioned as being fundamentally different from matter. While matter can be thought of as the result of Nature's automatic working, super-matter can be thought of as the result of a conscious will and cohesive wanting that causes a process of quantization by which the very fabric of matter is changed. As suggested in Section 10, such quantization is related to enhancing the functional-richness of matter. Imagine through a cohesive-wanting matter being able to become more flexible, more durable, more stretchable, light at will, heavy at will, change color at will, expand in size, contract in size, amongst infinite other functions and possibilities contained in Light at infinity.

Section 10 explored the micro-level process by which quantization changes the multi-dimensional space-time-energy-gravity fabric of material existence. This chapter takes a more macro-level view to summarize three distinct phases of matter: that of established-matter, matter-in-transition, and super-matter. In the process the requirement for an age of super-matter is made clearer. Note that this chapter leverages the mathematics developed in Chapter 3.6 for untransformed-organizations, organizations-in-transition, and transformed-organizations, with matter viewed as a particular instance of an organization. Hence the three 'phases' of matter are parallel to the three types of organizations considered in Chapter 3.6.

The Light-Space-Time Emergence equation (3.1.3) derived in Chapter 3.1, and reproduced here for convenience, will be leveraged to derive summary equations to distinguish between the different phases of matter. Reproducing the equation:

$$Emergence_{light-space-time} = \left\| \begin{matrix} \begin{bmatrix} C_\infty: [Pr, Po, K, H] \\ (\downarrow R_{C_K} = f(R_{C_\infty})) \\ C_K: [S_{Pr}, S_{Po}, S_K, S_H] \\ (\downarrow R_{C_N} = f(R_{C_K})) \\ C_N: f(S_{Pr} \; x \; S_{Po} \; x \; S_K \; x \; S_H) \\ (\downarrow R_{C_U} = f(R_{C_N})) \\ C_U: [P, V, M, C] \end{bmatrix}_{Light} & \begin{bmatrix} M_3 \to System_X \\ (\uparrow F \to I) \\ M_2 \to S_{System_X} \\ (\uparrow Sig \to F) \\ M_1 \to Sig_x \\ (\uparrow > P_x) \\ U \to x_U \end{bmatrix}_{Space} \\ \begin{bmatrix} M_3 : -\infty \leq t \leq \infty \\ \downarrow \\ M_2 : 0 \geq t > \infty \\ \downarrow \\ M_1 : 0 > t > \infty \\ \downarrow \\ U \to \begin{matrix} t \sim E_{Cell}; TC: M_3 \to U \\ t \sim E_{Human}; TC: U \to M_3 \end{matrix} \end{bmatrix}_{Time} & TC \to x_T \end{matrix} \right\|_{\langle x_U | x_T \rangle}$$

Simplifying each of the main matrices yielded (3.6.1), the simplified version of the Light-Space-Time Emergence equation, reproduced here for convenience:

$$Emergence_{light-space-time} = \left|[L][S][T]TC \to x_T\right|_{\langle x_U \mid x_T \rangle}$$

As explained by Neil Turok in his book The Universe Within (Turok, 2012), Euler's formula reproduced below, can be used to model many naturally occurring phenomena because of its sinusoidal oscillation between narrow bounds as x increases. Reproducing Euler's formula:

$$e^{ix} = \cos x + i \sin x$$

The sum of the squares of the ordinary and complex parts, on the right side of the equation, is one. In quantum theory this ensures that the probabilities for all possible outcomes add up to one. Hence this formula is useful when considering the macro-level effects of quantization of the fabric of material existence.

Further, in modeling matter a modified notation of the Schrodinger wavefunction as interpreted by Feynman is leveraged. This version features the integral sign, $\int$, meaning that all terms to the right of it have to be summed up for all space and time till the moment when the wavefunction is required to be known.

Summary Equations for Established-Matter

Hence, combining (3.6.1) with Feynman's interpretation of the Schrodinger wavefunction, with the Euler formula yields the following equations, 11.1.1-6 for matter:

$$\psi_{established-matter} = \left| \int e^{i \int ||[L][S][T]||} \right|_U$$

Eq. 11.1.1: Wavefunction for Established-Matter

(11.1.1) suggests that so long as the basis of matter is untransformed, specified by the U following the vertical-brackets, the outcome is not going to be any different from established matter, specified by $\psi_{established-matter}$.

Further, as specified by Equation 11.1.2, the Probability-View of Established-Matter, the very basis of matter is going to be either the untransformed physical (^{P}U), the untransformed vital (^{V}U), the untransformed mental (^{M}U), or the untransformed integral (^{I}U):

$$|\psi_{established-matter}|^2 = P_U^2 + V_U^2 + M_U^2 + I_U^2 = 1$$

Eq. 11.1.2: Probability-View for Established-Matter

As specified by (11.1.2) the probability that any of these bases will be leveraged in the creation of matter adds up to one. Note that the physical, the vital, the mental, and integral were discussed in Chapter 2.6 on Inherent Dynamics of a System.

Note also that another way of viewing established matter is by leveraging the non-wavefunction form for an untransformed organization (3.6.2) and adapting it to create Equation 11.1.3, Simplified Light-Space-Time Matrix Form Established-Matter:

$$Established-Matter = \left| |[L][S][T]TC{\rightarrow}x_T|_{\langle x_U | x_T \rangle} \right|_U$$

Eq. 11.1.3: Simplified Light-Space-Time Matrix Form for Established-Matter

Summary Equations for Matter-in-Transition

Equation 11.1.4, Wavefunction for Matter-in-Transition, suggests the mixed bases for matter, as specified by dynamics of both the untransformed (U) and the meta-levels (^{M}x), which therefore results in matter-in-transition.

$$\psi_{matter-in-transition} = \left| \int e^{i \int ||[L][S][T]||} \right|_{U\&M_x}$$

Eq. 11.1.4: Wavefunction for Matter-in-Transition

As specified by Equation 11.1.5, Probability-View for Matter-in-Transition, the probability that any of the untransformed (U) and transformed (T) bases will be leveraged in the creation of matter adds up to one.

$$|\psi_{matter-in-transition}|^2 = P_U^2 + P_T^2 + V_U^2 + V_T^2 + M_U^2 + M_T^2 + I_U^2 + I_T^2 = 1$$

Eq. 11.1.5: Probability View for Matter-in-Transition

Note also that another way of viewing matter in transition is by leveraging the non-wavefunction form for an organization in transition (3.6.3) and adapting it to create Equation 11.1.6, Simplified Light-Space-Time Matrix Form for Matter-in-Transition:

$$Matter-in-transition = \left| |[L][S][T]TC \rightarrow x_T|_{\langle x_U | x_T \rangle} \right|_{U\&M_x}$$

Eq. 11.1.6: Simplified Light-Space-Time Matrix Form for Matter-in-Transition

Summary Equations for Super-Matter

Equation 11.1.7, Wavefunction for Super-Matter, suggests some transformed bases for matter, as specified by dynamics of the meta-levels (^{M}x), which therefore results in super-matter.

$$\psi_{super-matter} = \left| \int e^{i \int ||[L][S][T]||} \right|_{M_x}$$

Eq. 11.1.7: Wavefunction for Super-Matter

As specified by Equation 11.1.8, Probability-View for Super-Matter, the probability that the transformed (T) bases will be leveraged in the creation of matter adds up to one.

$$|\psi_{super-matter}|^2 = P_T^2 + V_T^2 + M_T^2 + I_T^2 = 1$$

Eq. 11.1.8: Probability View for Super-Matter

Note also that another way of viewing super-matter is by leveraging the non-wavefunction form for a transformed organization (3.6.4) and adapting it to create Equation 11.1.9, Simplified Light-Space-Time Matrix Form for Super-Matter:

$$Super - Matter = \left| |[L][S][T]TC \rightarrow x_T|_{\langle x_U | x_T \rangle} \right|_{M_x}$$

Eq. 11.1.9: Simplified Light-Space-Time Matrix Form for Super-Matter

Chapter 11.2: Expansion of the Universe

In previous chapters it is suggested that a four-fold quantization is required for the fabric of material existence to change. Further in Chapters 10.1 and 10.2 it is suggested that a conscious and cohesive want for added functional-richness at the material level can trigger a process culminating in quantum-certainty in which such quantization occurs. In Chapter 11.1 a differentiation is made between three types of matter – ordinary-matter, matter-in-transition, and super-matter.

Development of functional richness precipitating in a change of matter can happen in either of these types of phases. In ordinary-matter it is automatic though through the coordination of Nature. In matter-in-transition there may be cohesive want for added functional richness that requires the overcoming of habitual patterns. But equally, material reality can be prolonged through habitual patterns hence broadening the bases for functional-enhancement to include both untransformed function and transformed function. In super-matter a cohesive wanting implies the consistent overcoming of habitual patterns, and the cumulative effect at the macro-level is that a foundation of super-matter is laid which then sets in place a virtuous cycle as suggested by the associated Schrodinger wavefunction for super-matter as explored in Chapter 11.1.

If the cumulative effect of overcoming habitual patterns increases the likelihood of super-matter being formed, this implies the increase too of four-fold quantization. Space, as the seeding ground of knowledge, will therefore also require expansion to continue to allow seeding to take place. But if space needs to expand in such a scenario, then the fact that scientists have observed that the universe is expanding may be related to this.

This chapter explores a possible link between the expansion of the universe and an increase in functional-richness that implies the expansion of space.

The entertainment of such an idea perhaps can be strengthened through consideration of the cosmological principle that suggests that the spatial distribution of matter in the universe is homogenous and isotropic when viewed on a threshold scale of 250 million light years.

The nine-year cosmic microwave background image created with the Wilkinson Microwave Anisotropy Probe (WMAP) depicts 13.77 Billion year old temperature differences that correspond to the seeds that became the galaxies (NASA-WMAP, 2014). This homogeneity in temperature difference in the early universe suggests too a process of homogenous four-fold quantization that set into motion the development of galaxies.

Complexity of Matter, Dark Matter, & Dark Energy

According to NASA only 5% of the matter in the universe is visible, while as much as 27% is 'dark matter'. It is the 5% of 'visible' matter though that forms the familiar stars, galaxies, or the atoms (NASA-darkmatter, 2016). But since, as explored in Chapter 2.7 on Qualified Determinism, the architectural sets at M_2 are infinite the question is why would all functions need to manifest in the same way?

As stated in an article 'Dark Energy: The Biggest Mystery in the Universe' (Panek, 2010), "Sight itself has blinded us to the Universe". It should be possible for other matter-based constructions to come into being based on the functionality they exist for. For instance, as already suggested there could be other fields that create other kinds of elementary particles, that create other kinds of atoms and that results in other kinds of structures not visible with our current instrumentation.

Recall (9.2.33) on complexity of matter in terms of the manifested sets:

$$Matter_{Complexity} \propto \left| MS_{System_{Pr}} \cup MS_{System_P} \cup MS_{System_K} \cup MS_{System_N} \right|$$

It is also possible to state an Equation, 11.2.1, relating complexity of matter in terms of visible and dark matter:

$$Matter_{Complexity} \propto [Visible\ Matter + Dark\ Matter]$$

Eq. 11.2.1: Complexity of Matter in Terms of Visible and Dark Matter

Further, as suggested by the previous discussions on the quantization of space in Section 1,3, and others, space is not empty but at the very least is a seeding ground for a variety of emergences. The notion of space having 'amazing' properties was, according to a report on 'dark energy' by the Harvard-Smithsonian Center for Astrophysics, first suggested by Einstein (Harvard-Smithsonian Center for Astrophysics, 2004). Quoting: "Einstein was the first person to realize that empty space is not nothingness. Space has amazing properties, many of which are just beginning to be understood." Einstein had suggested the existence of a 'dark energy' about 100 years ago (NASA-Supernova, 2001) as a property of space that caused the expansion of the universe. Dark Energy is estimated to comprise as much as 68% of the universe.

Reinterpretation of Cosmological Expansion-Contraction Dynamics

Keeping in mind the complexification of matter as it journeys from the field-level through the quantum-, atomic-, and cellular-levels, and beyond, as summarized in Figure 9.2.1 on the complexification of matter in terms of levels of implicit wholeness it may be possible to re-interpret the supposed expansion-contraction dynamics of cosmology in relation to the mathematical model presented in this book.

First, flipping the left and right sides of the equation (9.2.33) on $Matter_{Complexity}$ to yield Equation 11.2.2:

$$\left|MS_{System_{Pr}} \cup MS_{System_P} \cup MS_{System_K} \cup MS_{System_N}\right| \propto Matter_{Complexity}$$

Eq 11.2.2: Flipped Matter-Complexity Equation

This implies that the manifested-set, MS, is growing at a certain threshold level. Assuming this threshold level, $MS\_Growth_{Threshold}$, is a property of space related to dark energy it may be possible to restate the condition of cosmological expansion and contraction. Hence, so long as the MS is increasing at a certain rate that exceeds

$MS\_Growth_{Threshold}$ the level of dark energy is such that the acceleration of galaxies, $Acceleration_{Galaxies}$ exceeds the contracting force of gravity at the universal level, $Gravity_{Universe}$.

Therefore, as in Equations 11.2.3 and 11.2.4:

$$MS\_Growth_{Threshold}\text{:}\ Acceleration_{Galaxies} > Gravity_{Universe}$$

Eq. 11.2.3: Expansion of Universe Related to Manifest Set Growth Threshold

Conversely:

$$!MS\_Growth_{Threshold}\text{:}\ Acceleration_{Galaxies} < Gravity_{Universe}$$

Eq. 11.2.4: Contraction of Universe Related to Manifest Set Growth Threshold

The relation between $MS\_Growth_{Threshold}$ and Dark Energy is suggested by Equation 11.2.5:

$$MS\_Growth_{Threshold} \propto Dark\ Energy$$

Eq. 11.2.5: Relation Between Manifested Set Growth Threshold and Dark Energy

In this interpretation so long as more of the infinite set at M_2 results in the manifested-set, MS_{System}, the universe keeps growing.

Further if the relative distances between some of the levels as laid out in Figure 9.2.1 are considered a pattern emerges. The quantum level is dealing in distances of the order of 10^{-35}m. The cellular level is dealing with distances of the order of 10^{-6}m. The astronomical level is dealing with distances as large as 10^{26}m – the estimated size of the universe. It can be observed therefore that the cell is approximately in the center with roughly 30 orders of magnitude on either side.

Such symmetry that we are at now with the human being at the mid-point between two known material extremes, seems to suggest a cosmic balance of scales with human-will and other faculties such as awareness as a potential arbiter. Lack of will for continued functional-richness will in this view lead to a reversal in the Manifested Set due to a reversal in the quantization trend and consequently to a return of this universe to a state of collapse or unfulfilled absorption into the initiating Light. In that case it will be another possible Big Bang initiated through the interplay of multiple layers of Light, and a birth following such death as depicted by the ancient Ouroboros symbol of a dragon or snake eating its own tail.

On the other hand the will for continued functional-richness may lead to continuous quantization, space expansion, and a new Ouroboros of eternal return.

A New Ouroboros

Perhaps it is that if the expansion continues the apparent gaps between galaxies filled with extraordinarily high dark-energy becomes ripe for a series of inter-galactic-bangs. And perhaps such a series of inter-galactic-bangs in such a functionally-rich universe precipitates cycles of development based on other speeds of light, that in turn sets up a more conscious fusion of the multiple layers of existence so created, into one incredible Universe where an entirely new type of super-matter can form.

The new Ouroboros will signify a new type of endless creation where richness builds on richness and new types of super-matter on previous super-matter.

The human has in a certain sense been the result of 30 orders of microcosmic magnitude implicit to it. Fields, quanta, atoms, cells, the incredible four-fold emergences of properties codified in Light have all contributed to the emergence of the human. Now in a possible reversal of causality the human appears to be positioned to similarly influence and even structure the further development of the macrocosm by mastery of the microcosm.

Through will and other human faculties continued states of quantum-certainty resulting in four-fold quantization can occur. Such quantization changes the very nature of space, time, energy, and gravity and can alter the development of the 30 orders of macrocosmic magnitude.

Such audacity is only possible because the human is the latest emergence in a universe of Light where all is Light and all is of Light. Light in material form turning on itself can alter the very nature of what it is by what it is. Such is the tale of Ouroboros changing from swallowing its tail to entwining itself through a Cosmos in a never-ending spiral toward unforeseen flowerings.

SECTION 12: Birthing a New Ouroboros

The knowledge of light can be used practically to begin to change the nature of experience on multiple scales.

What is given is only the past. All the future has yet to be made and such creation is restricted only by imagination.

This Section suggests some exercises and reflections the nature of which can shift experienced reality:

- Healing the Body
- Fullness of Light
- Meditation on Four-Foldness of Emotion
- Shifting the World with Light
- Reflections on Creation

Chapter 12.1: Healing the Body

Our bodies are entirely a play of Light. Our bodies are made from cells and these cells are nothing other than a multi-layered structure emerging entirely from Light.

The first detectable layer is that of the electromagnetic field which structured as the range of waves, the gradient of energy, the ability to localize as particles, and moving at the speed c, is nothing other than a projection of Light's properties of Knowledge, Power, Service, and Harmony respectively. Quantum particles are localization of such an underlying field that similarly gathers into quarks, leptons, bosons, and the Higgs-boson, which are emergences of Knowledge, Power, Harmony, and Service respectively. Bosons themselves that connect quantum particles together to create atoms surface as Gluons, W and Z bosons, Photons, and the hypothetical Graviton. But these emergences are themselves that of Knowledge, Power, Service, and Harmony respectively. As a result quantum particles gather to create atoms, and all known atoms are similarly either of the s-shell or Power, p-shell or Knowledge, d-shell or Service, f-shell or Harmony categories. All molecules are made from these known atoms and therefore are nothing other combined functions of the infinite variation of the essential four properties of Light. All cells are made from molecules, and further, are themselves divided into Nucleic Acids, Polysaccharides, Lipids, and Proteins, which are themselves nothing other than emergences of Knowledge, Power, Harmony, and Service, which as we know are the properties of Light.

So our entire bodies are made from light and are nothing but light. An itch or a pain or a swelling or any other dysfunction is simply a result of the cells acting in their forgetfulness of their truth. Their truth is that they are concrete projections and crystallizations from an eternal and infinite world of Light and that they can partake in that reality. It does not necessarily mean that they themselves will become eternal and infinite, just yet, but it means that always they can allow some sense or even concreteness of eternity and infinity to enter into the dysfunctions and begin to heal them.

This is energy, or information, or more appropriately Light Medicine, and it is the panacea for any bodily ills. What is required is that the contracted time and space dynamics of pain or ill are allowed to open to the truth or Light or expansive timelessness and spacelessness that exist always behind it. The separation needs to be annulled by allowing the dysfunction to achieve fruition by connecting to the vastness of Light. For in its reality the pain or dysfunction is nothing other than a visceral urge to move away from something in which the separation is becoming more and more acute.

So, focus on an area in the body that is in dysfunction. See it in as much detail as you can – the tissues or muscles or bone or ligaments. See further that these are all made from cells. Now see these cells as nothing other than a play of light – the essential properties of light are there in every cell. But go deeper yet and see that all the molecules and atoms and quantum particles and the very field that everything emerges from are nothing other than the play of Light. Now know that Light can never be separated, and further that every smallest quantum particle is nothing than the Full Play of Light in an apparently small form. Now keep seeing this complete fullness of light in all the minuteness in the area of dysfunction. Keep doing this. Keep imagining that the tissue, the cells, the molecules, the atoms, the quantum particles, the very field from which all these emerge is flooded with the fullness of Light in its infinity and eternity and with its full play of properties and vast number of qualities. THAT is all there always and everything in the area of dysfunction is nothing other than selective functions that in their separation have forgotten that connection. Reconnect everything. Keep reconnecting it. See that this area is nothing other than that which has emerged from Light, and is nothing other than the full play of Light in a little space.

Chapter 12.2: Fullness of Light

Much of our time is spent in feelings, or thoughts, or urges that keep us occupied with smallness. We get entwined with the movement of separation often becoming one with it.

But if we can see that any of these movements has all the Fullness of Light behind them then something of that fullness can come into the smallness and that can alter the consequence of the smallness. For the fullness is everywhere and in everything already. The smallness is just here and now in what has ceased us. Being everywhere and in everything already the consequence of opening to it can be entirely different.

It is not that our little urge, or our desire, or our little thought will be guaranteed to be fulfilled because we are somehow opening to the fullness, but rather that the fullness, if we can open to it will make happen what is best from the point of view of the fullness of light. And living in that knowledge of the best possible happening, or living in the feeling of the best possible happening, allows the littleness or smallness that may have ceased us to open more to the fullness, and a virtuous cycle is then put in place.

The more we can practice this through the day, with all our smallnesses, the more will the largeness of light fill the smallness and allow a completely different dynamic to take place.

The Fullness of Light slowed down to c and allowed this matter-based creation - a complex crystallization of light – to come into being. But further the Fullness can be in the smallness, and can make the apparently small matter-based creations become full creations with access to oneness, and all the possibilities and potentialities resident in that Fullness.

To think, to feel, to sense, to will, to desire are great opportunities, great gifts even that each of us is always endowed with, a thousand times a day, if we use them as such. For being in their essence nothing but emergences of the properties of Light, and having all the Fullness of Light resident or behind each of these divisions, means that these divided dynamics that seem to lead us further from Fullness, can become means to lead us instead back into Fullness, or rather can become a passage by which the Fullness finds one more means or one more hold to manifest more of itself in these smaller infinitely diverse matter-based creations.

So thinking that the Fullness of Light is right here, and seeing that the Fullness of Light is right here, and willing and desiring that the Fullness of Light is right here, and loving that Fullness of Light so that it becomes more and more real right here is what will make the right here full of the Fullness of Light.

Chapter 12.3: Meditation on Four-Foldness of Emotion

Light is one. When light surfaces as a layer its components get separated and it is only the action of Love that bring about the unity in matter and life. This has to be kept in mind when meditating on emotions.

So, knowing that emotions are an emergence of Light's property of Harmony, it should be remembered that Light's other properties are probably close by and surfaced too in the emergence of the emotion, and bringing these into a balance or unity will help the emotion be resolved if it is a negative emotion. For an emotion usually has a trigger, and this may be some expectation of attachment caused by the way we think of ourselves or what is ours or what should be ours or how someone should have interacted with us. So there is this thought component that will accompany the emotion, and this thought component can be seen to be a surfacing of Light's property of Knowledge. But then the emotion will usually also surface somewhere in our body or cause discomfort somewhere in our body. And this location in the body can be thought of as a surfacing of Light's property of Presence. And finally the entire complex will have a charge to it, a kind of a power that makes us aware that something is not quite right. This surfacing of the charge can be thought of as an emergence of Light's property of Power.

So we see that the emotion has several aspects to it and these are always there and that these aspects are none other that the components or properties of Light. These aspects can continue out of balance and the negative emotion, its effect on the body, the thought that gives it legitimacy, and the charge it is creating can continue forever unless Love is applied to bring these components into a balanced unity.

The application of love would mean that the imbalance is adjusted so that if it were a restrictive thought that is perpetuating the negative experience that this restrictive thought is replaced by another kind of more expansive thought. It would mean that the resulting charge is acknowledged and the impact on the body is acknowledged and holding the four-fold reconfigured emotion-complex in non-judgment, something else is allowed to surface instead.

Such a practice will likely take repetition before results are seen.

Chapter 12.4: Shifting the World with light

As our muscle for imagining, thinking, seeing, feeling, urging, willing, loving Fullness of Light grows, it becomes easier to see, feel, imagine, urge, will, love Fullness of Light in circumstances and situations that are apparently disconnected from us. We could be sitting in San Francisco, and engage in such Light exercises in or on other parts of the world entirely.

But starting small, and with the self, of which there are already so many opportunities, and then moving to larger and more distant spheres is a good way to proceed. Because in so doing it is as though a kind of "lubrication" sets in, or the way Light flows in and around the small self has begun to change. Then it becomes easier to exercise the Light in the other, more distant, larger spheres.

So for example, pick a part of the world where you know there are parties in conflict because they are embedded in different points of view, for instance. Now, even if you do not know these different points of view, or do not understand the subtleties they are expressing, still see that the thoughts, the feelings, the actions that are being exercised there are a surface rendering of the play of light, and also that these dynamics are separated from the Continent of Light and therefore often in chaotic motion.

So see first, or will, or imagine that all the Fullness of Light is right there and in fact everything, all the smallnesses are just emergences of that or projections of that, and THAT is right there informing everything. Imagine, or see, or love the Light in such a way that it surrounds the entire area and let it flow into every aspect and let every aspect be filled with it, and let the whole area be filled with it, so that there is nothing there but Light.

Do this in such a way that there is no personal will, or urge, or thought favoring one outcome or resolution over another. Just do it so that the Light itself, which is already in everything, may resolve the situation from its own point of view. If you know what some of the players in your area of focus look like, imagine them too in that sphere of Light. See them, every part of them down to the smallest quantum particles being nothing other than a play of light, and offer again the whole situation to the Light that it may be best taken up by the Light from its point of view.

If you have a bias to imagine such things, rather than see such things, or will such things, whatever your bias that is fine. Follow the emerging or emergent property of Light that most easily comes to you. But also then as you feel the flow of Light, use the other faculties or the fullness of all the properties as well as you are able to in the exercise of shifting this other part of the world. So use your thought of Light, your will for Light, your urge for Light, your sense of Light, your seeing of Light, your feelings for Light, your hearing of Light, your love for Light, to all become active in that sphere of your focus. If possible keep doing it until something in you feels that it is enough for now.

Chapter 12.5: Reflections on Creation

Imagine somewhere Light existing in a native state of unimaginable power. In that state Light travels infinitely fast and there is nowhere that it is not at, at the very moment. In fact, because it is so fast it is everywhere at once, and therefore there is only the Now. There is no past, present, or future. Since it is everywhere at once it knows everything at once, and since nothing can be that is not of its nature, it is all-powerful in that state. Further, anything that may arise is bound to everything else in Light, and therefore too there is a Light-based harmony everywhere.

So that light is present everywhere, is all-powerful, knows everything, and creates the ultimate harmony.

Now imagine that this Light, having all possibility present in itself, seeks to project at least some of these out side of itself. It seeks to create based on its own nature. In its all-knowledge it decides that the way it will project something if itself, is to project versions of itself at different speeds.

Darkness at Zero

At one extreme it projects out a version of itself that has zero speed. But traveling at zero speed means that it cannot go anywhere – its light therefore cannot travel and what results is utter darkness. The Light at zero cannot know anything around it, cannot have any power outside of its infinitesimally small point, cannot embrace anything besides its smallness, and cannot be present anywhere except at the completely fragmented spot it is currently at. So the creation that results is the opposite of the creation that existed when Light was infinitely fast. Instead of knowledge there is utter ignorance, instead of power there is complete weakness, instead of harmony, there is chaos, and instead of an all-presence there is a complete fragmentation.
It is this four-fold state of ignorance-weakness-chaos-fragmentation that also has an influence on matter, and holds it in a contrary embrace. Battle between light and dark therefore takes place in matter, where even though matter is implicitly Light and can open and even change as per Light, yet it is also always under the influence of darkness as well.

But all the while even this complete darkness and ignorance and weakness and fragmentation and chaos are nothing other than a play of Light. Everything that Light is at infinity, it still is at zero, except that it is hidden away in infinitesimally small points.

So even when there are these four dismal states yet these can be seen as a veil of Light and in that knowledge there may still be some liberation that even behind the apparent worst there is still utter Light.

Packets at c

Now what when light projects itself out of its infinity at other speeds? At c this wonderful material universe we are aware of is created. For in the act of slowing down any possibility that exists in Light will need to 'accumulate' itself before it can be expressed. So the whole notion of packets of Light comes into being.

And since Light has infinite possibility in itself, arranged in four master-sets of knowledge, harmony, presence, and power, the infinite possibility expresses itself in un-ending combinations of elements derived from these four master-sets. So there are an infinite number of seeds that are created. These seeds need to be organized to exist, and that is what Space is. Space is the organization of an infinite number of seeds.

But the seeds coming into being in a layer of Light at c will only be able to express themselves after some accumulation and so there will also be a process of packet-creation or quantization in the expression of seeds.

But further, these seeds will need to go through some process of maturation to express what is in them, and that is why Time is needed. Time is an arrangement to allow maturation at the level of matter to occur. And since this Time is happening at a layer where Light travels at c, there will have to similarly be some process of packet-creation or quantization related to the way Time expresses itself.

So what is in Space and expresses itself in Time will materialize at some point as the essential energy of its being accumulates. But occurring at a layer where Light is traveling at c there will have to similarly be a process by which Energy accumulates in packets to become Matter as it crosses some threshold.

And then, the seed will require a 'system' or ecosystem in which to flourish, and so will have relationships with other seeds or create a particular nature of relationships within its ecosystem and that is what Gravity is. But Gravity too, being at a layer where Light is traveling at c will require a packet-like or quantization process to express itself.

Matter at c

Matter then is not just made up of layers of Light's emergences of quantum particles and atoms, but also is the result of space-time-energy-gravity "packetization" or quantization. This means that matter is not fixed but is variable and that the meaning or reality of matter can continually change. For matter is then equally dependent on seeds and so can be changed as new seeds are created or envisioned or precipitated.

Since seeds are combined from four sets that are essentially sets of function, anything that can influence function or change function, can therefore influence the form of matter.

Matter at c then can be highly variable becoming flexible, or durable, or stretchable, or light-able, or expandable. There are no limits to what can be done with matter. The only limit is in the imagination.

Variable Speeds of Light

Light can move at an infinite speed creating a reality of all-power, all-knowledge, all-presence, and all-harmony; it can "move" at speed zero creating a reality of utter darkness; and it can move at c creating the material reality we are familiar with and have arisen in.

But Light can travel at other speeds too and it is the realities created at some of these other speeds that allows a precipitation of the infinite possibility implicit at the layer where Light is traveling infinitely fast to the immense diversity that is apparent in the material layer.

When light travels at a speed in-between c and infinity, but yet closer to c, it can be envisioned that there will be a point where light is fast enough that there cannot be a substantial accumulation of energy to create matter, but an accumulation enough to create waves. These waves are then in essence spread over vast distances instantaneously, and holding something of the infinite variety or information in light, can preside over diverse development at the material level.

So too, it can be envisioned that when light travels at a speed somewhere between c and infinity, but closer to infinity, light can "travel" in immense blocks.

Self as an equation that can change the mathematics of creation

Think of the blocks as the elements of the four sets, spreading their influence through vastness. Think of the waves as combination of these elements, creating unique seeds as it were, that then preside over material possibility at the layer where light travels at c.

So if any individuality expresses itself in apparent uniqueness, yet it is ONE with Light, and has all of Light implicit in it. This then is the great secret of the self – that implicit in self is all of Light ever-present. It is the same equation that expresses everything, that is also expressed as the self. So each self is connected to everything, and it is this dynamic that can give utter power to even shape apparent emptiness in space or the seemingly infinite distances between galaxies. The self is a mathematical-key, a mysterious variable, that turning on itself, can change the equilibrium of this mathematically dynamic creation.

It is this secret that will usher in a New Ouroboros.

REFERENCES

1. Arabatzis, T. 2006. Representing Electrons: A Biological Approach to Theoretical Entities. University of Chicago. Chicago
2. Arkani-Hamed, Nima; Bourjaily, Jacob L.; Cachazo, Freddy; Goncharov, Alexander B.; Postnikov, Alexander; Trnka, Jaroslav (2012). "Scattering Amplitudes and the Positive Grassmannian". arXiv:1212.5605
3. Bar-Yam, Y. 2016. From big data to important information. Complexity, 21: 73–98. doi:10.1002/cplx.21785
4. Briggs, J. 1992. Fractals: The Patterns of Chaos. Simon & Schuster: New York.
5. Chown, M. 1990. Can Photons Travel Faster Than Light? New Scientist 126(1711)
6. Cottingham, W & Greenwood, D. 2007. An Introduction to the Standard Model of Particle Physics. Cambridge University Press. Cambridge
7. De Broglie, L. 1929. The Wave Nature of the Electron. Nobel Lecture. http://www.nobelprize.org/nobel_prizes/physics/laureates/1929/broglie-lecture.pdf
8. Deep Order Mathematics Videos. 2016. Deep Order Technologies. http://www.deepordertechnologies.com/new-index
9. Diamond, J. 2005. Collapse: How Societies Choose to Fail or Succeed. Viking Books: New York
10. Einstein, A. 1995. Relativity: The Special and General Theory. New York: Broadway Books.
11. Feynman, RP. 1985. QED The Strange Theory of Light and Matter. New Jersey: Princeton University Press
12. Frost, R. 2005. Elliott Wave Principle: Key to Market Behavior. New Classics Library: Georgia
13. Goodsell, David. 2010. The Machinery of Life. New York: Springer
14. Gottlieb, M. 2013. The Feynman Lectures on Physics: III. California Institute of Technology. http://www.feynmanlectures.caltech.edu/III_16.html
15. Gray, T. 2009. The Elements: A Visual Exploration of Every Known Atom in the Universe. Black Dog & Levental Publishers. New York.

16. Gubser, S. 2010. The Little Book of String Theory. Princeton University Press
17. Harvard-Edu. 2014. http://news.harvard.edu/gazette/story/2014/04/harvard-to-sign-on-to-united-nations-supported-principles-for-responsible-investment/
18. Harvard-Smithsonian Center for Astrophysics. 2004. https://www.cfa.harvard.edu/seuforum/de_whatmight.htm#
19. Hawking, Stephen. 1988. A Brief History of Time. New York: Bantam Books
20. Heiserman, D. 1991. Exploring Chemical Elements and their Compounds. McGraw-Hill. New York.
21. Hope, Chris; Fowler, Stephen J. (2007). "A Critical Review of Sustainable Business Indices and Their Impact". Journal of Business Ethics. Vol. 76. S. 243–252. Springer: New York
22. Hoverstadt, P. 2008. The Fractal Organization: Creating Sustainable Organizations with the Viable System Model. Chichester: John Wiley & Sons
23. Hyperphysics. 2016. Department of Physics and Astronomy. Georgia State University. http://hyperphysics.phy-astr.gsu.edu/hbase/mod3.html#c1
24. Isaacson, W. 2008. Einstein: His Life and Universe. Simon and Schuster. New York.
25. Jeans, J. 1932. The Mysterious Universe. Cambridge University Press.
26. Jepsen, K. 2013. Everything is Made of Fields. Symmetry. Joint Publication by Fermilab/SLAC. http://www.symmetrymagazine.org/article/july-2013/real-talk-everything-is-made-of-fields
27. Johnson, S. 2010. Where Good Ideas Come From: The Natural History of Innovation. Riverhead Books: New York
28. Kaufmann, S. 1995. At Home in the Universe. New York: Oxford University Press.
29. Kaufmann, S. 2003. The Adjacent Possible, Edge. https://www.edge.org/conversation/the-adjacent-possible
30. Laszlo, E. 2014. The Self-Actualizing Cosmos: The Akasha Revolution in Science and Human Consciousness. Inner Traditions: Vermont
31. Logue, A. 2008. Socially Responsible Investing for Dummies. Wiley Publishing: New Jersey
32. Lorentz, H.A. 1925. The Science of Nature. Vol. 25, p 1008. Springer

33. Malik, P. 2009. Connecting Inner Power with Global Change: The Fractal Ladder. New Delhi: Sage Publications
34. Malik, P. 2015. The Fractal Organization. New Delhi: Sage
35. Mandelbrot, B, Hudson, R. 2006. The Misbehavior of Markets: A Fractal View of Risk, Ruin, and Reward. New York: Basic Books
36. Mora, C, Tittensor, D, Adl, S, Simpson, A, Worm, B. 2011. How Many Species Are There on Earth and in the Ocean? Plos.org. http://journals.plos.org/plosbiology/article?id=10.1371/journal.pbio.1001127#abstract0
37. NASA-darkmatter. 2016. http://science.nasa.gov/astrophysics/focus-areas/what-is-dark-energy/
38. NASA-supernova. 2001. News Release Number: STScI-2001-09: "Blast from the Past: Farthest Supernova Ever Seen Sheds Light on Dark Universe". http://hubblesite.org/newscenter/archive/releases/2001/09/
39. NASA-WMAP, 2014. https://map.gsfc.nasa.gov/media/121238/index.html
40. Ogburn, W. Tomas, D. 1922. Are Inventions Inevitable? A Note on Social Evolution. Political Science Quarterly 37, no.1
41. Olive, K.A et al. 2014. Particle Data Group. Chin. Phys. C, 38, 090001.
42. Openshaw, J. 2015. http://www.marketwatch.com/story/socially-responsible-investing-has-beaten-the-sp-500-for-decades-2015-05-21
43. Panek,R. 2010. Dark Energy: The Biggest Mystery in the Universe, Smithsonian Magazine. http://www.smithsonianmag.com/science-nature/dark-energy-the-biggest-mystery-in-the-universe-9482130/?no-ist
44. Particle Data Group. 2015. Lawrence Berkeley National Laboratory. http://www.cpepphysics.org/main_universe/universe.html
45. Patterson, K, Grenny, J. Crucial Conversations: Tools for Talking when Stakes are High. 2011. McGraw-Hill: New York.
46. Pauli, W. 1964. Nobel Lectures, Physics 1942 – 1962. Elsevier Publishing Company. Amsterdam.
47. Pearson, R. 1997. "Frontier Perspectives", Journal of the Center for Frontier Sciences at Temple University.

Spring/Summer 1997, Volume 6, Number 2, ISBN: 1062-4767.

48. Perkowitz, S. 2011. Slow Light. London: Imperial College Press
49. Planck, M. 1933. Where is Science Going? Ox Bow Press. Connecticut.
50. Platzek, B., Pretorius L., & Winzker, D, 2014, The Vital Entrepreneurial Learning Organization: A Corporate Mindset for Entrepreneurial Change Management, International Journal of Innovation and Technology Management, Vol. 11, No. 6 (2014) 1450044, World Scientific Publishing Company
51. Portugali, J., 2012. *Self-organization and the city*. New York: Springer Science & Business Media.
52. Prigogine, I. 1977. Time, Structure, and Fluctuations. *Nobelprize.org*. Nobel Media AB 2014. Web. 5 Mar 2016. <http://www.nobelprize.org/nobel_prizes/chemistry/laureates/1977/prigogine-lecture.html>
53. Salkind, N. 2007. Encyclopedia of Measurement and Statistics. Thousand Oaks: Sage Publications.
54. Snyder, M. 2010. Stanford Medicine. http://med.stanford.edu/news/all-news/2010/03/what-makes-you-unique-not-genes-so-much-as-surrounding-sequences-says-stanford-study.html#.html
55. Sri Aurobindo. 1971. Social and Political Thought. Sri Aurobindo Ashram Press: Pondicherry
56. Stewart, Ian. 2012. In Pursuit of the Unknown. Basic Books. New York.
57. Swimme, B. 2001. The Universe is a Green Dragon. Bear & Company: Rochester
58. Smith, J, Szathmary, E. 1995. The Major Transitions in Evolution. Oxford: Oxford University Press
59. Thomas, D, Brown, J. 2009. Why Virtual Worlds Can Matter. International Journal of Learning and Media 2009 1:1, 37-49
60. Toynbee, A. 1961. A Study of History, Volumes I – XII. Oxford University Press: Oxford
61. Turok, N. 2012. The Universe Within. Anasi Press: Ontario
62. Tweed, M. 2003. Essential Elements: Atoms, Quarks, and the Periodic Table. Walker & Copmany. New York.
63. Usó -Doménech, J, Nescolarde-Selva, J, Lloret-Climent, M, Fan, M. 2016. Semiotic open complex systems: Processes

and behaviors. Complexity, 21: 388–396. doi:10.1002/cplx.21817
64. Whitaker, A._2006. Einstein, Bohr and the Quantum Dilemma: From Quantum Theory to Quantum Information. Cambridge: Cambridge University Press
65. Wilczek, F. 2016. A Beautiful Question: Finding Nature's Deep Design. New York: Penguin Books
66. Willis, A. 2003. The Role of the Global Reporting Initiative's Sustainability Reporting Guidelines in the Social Screening of Investment. Journal of Business Ethics. Volume 43. Springer: New York
67. Wolchover, N. 2013. A Jewel at the Heart of Quantum Physics. Quanta Magazine . https://www.quantamagazine.org/20130917-a-jewel-at-the-heart-of-quantum-physics/
68. Wright, R. 2009. Evolution of Compassion. https://www.ted.com/talks/robert_wright_the_evolution_of_compassion/transcript?language=en. TED 2009.
69. Yates, F.E. 2012. *Self-organizing systems: The emergence of order*. New York: Springer Science & Business Media.

Genesis of *Cosmology of Light* Series

The Cosmology of Light series was envisioned as a series of six books. Each book is independent and complete in itself. Yet there is some overlap between the books, especially in Books 2 – 6, which contain many common mathematical elements.

The series of six was laid out at the beginning of the writing journey, based on essentially independent ideas that surfaced during my doctoral work on the mathematics of innovation in complex adaptive systems. That vision remained intact through the journey, and materials that overlap in subsequent books has in fact generally been worked out to a great detail, while still keeping the previous materials in earlier books complete for its purpose.

Brief Summary of Books

Book 1, A Story of Light: A Simple Exploration of the Creation and Dynamics of this Universe and Others, was envisioned to contain the core ideas about light that highlight the essential place of Light in the creation. But further this book also sought to express how everything that is, is nothing but Light. This book was just about the ideas, saving the mathematics for subsequent books.

Book 2, Oceans of Innovation: The Mathematical Heart of Complex Systems, focused on the essential mathematics required to think about the implicit creativity or innovativeness existing in any natural or evolving system. The mathematical model is built based on multiple speeds of light, each of which creates a simultaneous, independent reality. It is proposed that these realities interact and mathematics for this is proposed too.

Book 3, Emergence: A Mathematical Journey from the Big Bang to Sustainable Global Civilization, sought to apply some of the mathematics developed in Book 2, to the evolving creation, essentially highlighting that everything that has emerged can be expressed using the same set of mathematical equations. This has to be in a universe created from one source of Light.

Book 4, Quantum Certainty: A Mathematics of Natural and Sustainable Human History, focuses on the state of 'quantum certainty' by which history is made. The quantum certain state is part of the system mathematics that had been derived to explain the essential innovativeness of systems based on Light. Quantum certainty is a state that reinforces the notion of reality being one, with the 'observer' having an influence on what possibility precipitates based on Will or Cohesive Want.

Book 5, Super-Matter: Functional Richness in an Expanding Universe, suggests a relationship between matter, functional richness, and an expanding universe. It builds off Book 4, and contains most things explored in Book 4. Yet its focus is different and it extends the mathematical framework to the notion of 'super-matter' and an 'expanding universe'.

Book 6, Cosmology of Light: A Mathematical Integration of Matter, Life, History & Civilization, Universe, and Self, contains most ideas in each of the previous 5 books, but develops the mathematics further and is more comprehensive in its scope. It arrives at a more integrated picture of cosmos where self may be seen as a mathematical emergence with all the power of the cosmos' mathematics contained in it.

Author's Note to Book 1: A Story of Light – A Simple Exploration of the Creation and Dynamics of this Universe and Others

The journey to explore Light began in an unexpected place.

A Professor in South Africa, Dr. Dietmar Winzker, read a book of mine some years ago – Connecting Inner Power with Global Change: The Fractal Ladder. He wrote me a 16-page letter detailing his assessment of the book. To say the least I was thoroughly impressed that anyone would take the time to study the book in such detail, and then go on to very frankly lay out what they were in agreement with, and what not. Dr. Winzker's passion and commitment led to further dialog between us and in course of time he suggested I do a doctoral program from University of Pretoria in South Africa, where he had himself completed the first of his two doctoral degrees. He subsequently introduced me to Professor Dr. Leon Pretorius, his own Ph.D. supervisor, who was soon to also become my supervisor.

My research interest was in the area of Complex Adaptive Systems, and specifically in the mathematics of innovation of Complex Adaptive Systems. Simply saying, what captured my imagination was the question of whether a common mathematical framework, based on trends observed in time and subsequent insights into the structure of space, could explain the dynamics of innovation, regardless of the scale of the complex adaptive system, be it the size of a mere cell, or the entire universe.

Dr. Pretorius was an astute supervisor and led me through an essentially Socratic approach. He would ask question after question and let me discover my own answers. Since the field of Complex Adaptive Systems is essentially multi-disciplinary though, I found that I had to research a chain of adjacent areas in order to answer some of the questions posed. Pretty soon I had explored 25 areas ranging from Schrodinger's and Heisenberg's quantum equations, to the structure of the Periodic Table, to molecular plans in the human cell, to the stability of enduring civilizations, to the existence of dark matter and energy and their possible relation to the expansion and contraction of a universe.

In the course of this investigation the math of innovation in CAS had expanded to include over 225 equations, many of which were instances of application in the specific areas I had examined, and seeing that my enthusiasm to explore and connect seemingly distinct areas knew few bounds, Dr. Pretorius finally stopped me from exploring any more areas, especially since my thesis was fast becoming "daunting from a reviewer perspective" to paraphrase one of the potential examiners. The good news is that by the time he did this I had already happened on the mysteries of light.

Light, I found, was this extraordinary phenomenon, and in the mathematical framework that emerged held a foundational place such that everything in the universe could be thought to exist because of light. Light was the Holy Grail to arriving at a unified Theory of Everything, providing the clue to a Unified Field Equation as it were. Having glimpsed this I wanted to capture in as simple a language as possible the discoveries that continue to fill me with delight because of their simplicity in offering an interpretation for the creation and dynamics of a universe.

I was inspired by a non-technical book in plain English, written by Albert Einstein on his Theory of Special and General Relativity, and thought that if he could explain to a layman something as complex as Relativity, then I must be able to do the same on my interpretation of how everything emerges from Light. Hence I took it on my self to write this book that would not refer to a single equation, and would try and explain any complex phenomenon in simple terms. Necessarily this book is short with a key idea being more or less explained on a single page, and with the next key idea being dependent on the ideas already explored on previous pages. I have sought to cover the essential areas from my doctoral work, with the idea being that a reader with an excessively scientific or mathematical curiosity or mind could subsequently refer to the original dissertation if they so chose to. Details to access this appear later in the book.

Note that the nature of this work being more of that of the recognition and connection of pattern from area to area there will undoubtedly be instances where the details I attribute to parts of such patterns may require further thought or correction. But I hope the reader may begin to glimpse something that I glimpsed, and that these insights may be a constant source of joy and inspiration. If this is the case I further invite the reader to continue in creative-community with me and list some options for this at the end of the book.

Author's Note to Book 2: Oceans of Innovation – The Mathematical Heart of Complex Systems

We live in an infinitely innovative world. It is part of the very fabric of our existence. Being that this ability to innovate can express itself in massively different ways there are surely therefore a massive number of different approaches to express and frame innovation.

Here is one approach to frame such innovation. It has both an artistic and scientific dimension to it.

Artistic through necessity - in that when one tries to enter into the reality of what innovation is there is no expression that can fully represent it, and hence art seems at least to capture the current surge of such a contemplation. But the medium of the art here is mathematical since for me that alone seemed to capture the richness implicit in this de facto aspect of reality.

The mathematics expressed here is therefore the art. And this will be abundantly apparent in the attempt, for instance, to capture the reality of a point-instant infinite replete with all possibility. Necessarily the mathematical notation and the resulting mathematics had to expand beyond current norms in mathematics. This may raise eyebrows amongst some mathematicians. But as with my invitation in other recent writings I would welcome such individuals into collaboration to help capture perhaps in more robust ways what the artistic-mathematics here is seeking to express.

The mathematics expressed here is also a science. This is so because it proposes a structure and process to innovation that can be corroborated through observation and experiment. In its scientific aspect this book starts with a derivation of some subjective qualities of light that is proposed to set up objective reality. There is a four-foldness that is seen in this analysis that becomes the basis of the entire mathematics through the rest of the book. Hence the development of the mathematics primarily proceeds through induction, subsequently employing deduction to derive a representative set of mathematical operators that must be true of any complex system.

This book is therefore an artistic-mathematical contemplation of the aspect of innovation-reality that exists in every point-instant of any complex system. To make the most of it my suggestion is to suspend any judgment and enter into the unfolding contemplation to the degree possible. In this way you may see as I have seen and this may enrich in some way your own point of view.

Author's Note to Book 3: Emergence – A Mathematical Journey from the Big Bang to Sustainable Global Civilization

The point of this book is to illustrate a ubiquitous four-fold functionality that exists in the structuring of this universe. What has emerged through the process of time starting with a Big Bang to modern Civilization is not random but is ordered around this four-fold functionality of Presence, Knowledge, Harmony, and Service.

These core functional properties are observed in the nature of reality as set up by light itself. The process of the Big Bang, through the slowing down of light, and the inter-relation between implicit and projected properties of light, created a seed-equation that has subsequently propelled the process of emergence through numerous iterations.

These properties subsequently animate several layers of emergence starting from the 'field' level, through the 'quantum particle' level, the level of 'atoms', 'cells' and even 'civilizations'.

This book will present a mathematics that elaborates the relationship, the four-fold functional relationship to be precise, which exists between one layer of emergence and the next. As such the expression of the universe will be seen as unanimous with the expression of an evolving four-fold functional richness.

The notion of emergence presented here should be coupled with the fundamentally different notion of emergence commonly thought to be true of complex adaptive systems in general. In the latter, emergence is seen as unpredictable in that given the reality of quantum particles, say, it is impossible to predict that a fundamentally different and foundational physical form, that of atoms, would surface. Similarly it is impossible from just observing the dynamics of atoms that something like cells would emerge, or at a later stage even something as complex as civilization.

By coupling the two notions of emergence form can be seen as a 'switch' of sorts, or a receptacle, by which certain functionality can manifest. In this view the properties of Presence, Knowledge, Harmony, and Service take on a different meaning and put us face-to-face with a vast Complex Adaptive System that is our universe.

By examining the mathematics created from the Big Bang seed-equation, this book suggests that there are mathematical equations that need to be fulfilled for a sustainable global civilization to successfully emerge. Further, this book suggests an algorithm, also based on the Big Bang seed-equation, that ties together or rather propels the emergence of each subsequent and more complex layer.

Note that there are details in each of the layers that scientists and specialists – whether those focused on optics, or astrophysics, or quantum physics, or chemistry, or cell biology, or anthropology, or history, amongst other areas - may find incomplete or even misplaced. To these individuals I extend a special invitation to collaborate on working towards a more comprehensive description.

Author's Note to Book 4: Quantum Certainty – A Mathematics of Natural and Sustainable Human History

The world of quantum physics has been described as weird or absurd. In it are found phenomena such as dual wave-particle nature, superposition, quantum tunneling, entanglement, going backward in time, quantum teleportation, quantum uncertainty, amongst others, that appear very contrary to phenomena that we are used to in the visible day-to-day world.

This book presents a mathematical model, generated from properties of light, which proposes a context for interpreting quantum phenomena. Essentially, quantum phenomena are positioned as being at the interface between realities set up by different speeds of light.

As per the mathematical model presented here the quantum weirdness is not really weird and is a natural outcome of considering reality at the interface of different light-based worlds. While it is perhaps easier to have become aware of such phenomena at the micro-level because of the experimental rigor and attention to detail that has accompanied the study of quantum particles, it is proposed that such weirdness can under the right conditions exist also at the macro-level. The weirdness is positioned as appearing "weird" primarily because models of reality are force-fitted into a single layer of existence, when in fact it must be a complex of multiple interacting layers as suggested by the possibility of simultaneous and even interpenetrating realities so set up by multiple speeds of light.

Hence this book will also explore the conditions is which similar macro-level weirdness may be more easily observed or even created. Namely, in a culminating movement, this book will hypothesize that history, whether natural or human, is the result of sustained cohesive moments or activation-states, when the multiple-layer structure accessed through quanta aligns so as to create certainty, or more precisely quantum-certainty, in outcome. In this point of view such direction-setting dynamics has to be at the quantum-level because it is only through a process of coordinated micro-level space, time, energy, and gravity quantization that the very fabric of existence and possibility can change.

This book therefore proposes a process to create sustainable history that involves entering into an activation-state. Such an activation-state opens a quantum-window into layers of reality formed by light moving at different speeds. Collective meta-functions, whether of plants, animals, or humans exist in these layers and it is the enhancing or creation of such a meta-function that sets into motion a precise process of space, time, energy, and gravity quantization, by which the nature of reality can be changed. As such there is no uncertainty at these decisive moments of history, but a quantum-certainty that is worked out by the mathematical possibilities present at the time, and as elaborated by the mathematical model presented here.

Note that this book is the fourth in a series on the Cosmology of Light. Those who have read the previous books will find some overlap in the initial sections that explore the basics and mathematics of light.

Author's Note to Book 5: Super-Matter – Functional Richness in an Expanding Universe

We live at an incredible time in an incredible universe.

Through a play of Light its codified properties project out to create layer upon layer of fourfold complexity. Space, Time, Gravity, and Energy are perhaps the first layer of Light's codified properties and emerge as fundamental to cosmic dynamics to-date. Fields such as the electromagnetic field form a second layer. Then quantum particles, atoms, cells, and all the complexity of Natural and Human history form subsequent layers, all the while continuing to materialize the inherent fourfold-richness inherent in Light.

The fate of such a continued fourfold development leading to a foundation of super-matter now lies in our hands. Perfectly positioned between two apparent extremes, the human lies between approximately 30 orders of magnitude of micro- and macrocosm on either side that they are an intimate part of. All the depth and mystery of the microcosm, starting from the quantum-levels lies already implicit within the human. Perhaps even it can be said that the human is the result of all the emergent dynamics of these 30 orders of magnitude.

But now the scales of causality are reversing and the question is whether the human can similarly shape the further development of the macrocosm perhaps through mastery of the microcosm. Such audacity is only possible because all is Light and all is of Light. There is an intimate and implicit connection between apparent separativity regardless of scale and for the first time in history Nature appears to have strongly separated levers such as Will, awareness, and complex thought, amongst other faculties to a creature, and to exercise these to intimately structure and become a living part of the macrocosm is the challenge that lies before us.

The scales of a quantum-cosmic balance, therefore, depend on the extent to which a continued and cohesive space-time-energy-gravity quantization, fuelled by such will and other human faculties becomes operative. Such quantization spurred by the need for functional-richness results in super-matter. But further, such quantization may be intimately related to the energy known by cosmologists as 'dark energy' that continues to fuel the expansion of the universe.

In such an expansion the once recurring Big Bang, sign of the tail-biting Ouroboros in the throes of a constant cycle of destruction and rebirth, potentially yields to a new series of inter-galactic-bangs or other such creative mechanism, and sets into motion an endless development with super-matter building on super-matter.

Invitation to Continue the Exploration

If the ideas in this treatise appeal to you, or inspire you, or capture your imagination I invite you to creatively contribute in one or more of a number of ways. Here I list a few possibilities as starters:

- The treatise can be enriched with drawing and illustrations. These can be for children or adults and would create stand-alone books
- The treatise can be translated into different languages to share insights with larger audiences
- Further thinking and practical experimentation to do with any of the sciences referred to here whether to do with astrophysics, quantum physics, electromagnetic theory, optics, chemistry, biology, material science, nanotechnology, or any other fields or areas related to these. Working with active scientists and academicians in these fields will no doubt prove invaluable
- Further thinking and elaboration to do with the function-based mathematics referred to here. Working with real mathematicians will no doubt enrich the effort
- Further exploration into psychology of individuals as it relates to the expression of four-foldness of light
- Further exploration with organizational development practitioners to work through the implication of applying the light-framework explored here in practical terms
- Development of software tools to elaborate on one or more aspects covered in the treatise
- Construction of other meditations for healing at different levels or actualizing of potential that you would like to share
- Creation of workshops, seminars, conferences, living-room meetings to do with Light as it is explored here
- Creation of certifications for to-be light-coaches and light-facilitators
- Distribution of this content to people who would be interested in this material from a scientific, mathematical, organizational, psychological, or general perspective
- Translation of ideas here into a movie or expression onto other creative media

Please contact me if you would like to pursue or collaborate on any of these or any other idea at: Pravir.malik (at) deepordertechnologies.com

Made in United States
Orlando, FL
04 November 2026

90204435R00162